THE BIG NICKEL
Inco at home and abroad

THE BIG NICKEL
Inco at Home and Abroad

Jamie Swift and
The Development Education Centre

Foreword by Dave Patterson

Illustrated by Jaffé

© 1977 Between the Lines

Typeset and printed in Canada by union labour.

Published by: Between the Lines
97 Victoria Street North,
Kitchener, Ontario.

All graphics by Jaffé
Photographs: page 8 — Sudbury Historical Society
page 61 — *Sudbury Star*
page 121 — top — Sudbury Historical Society
— bottom — *Sudbury Star*

ISBN 0-919946-06-2 bd.   ISBN 0-919946-05-4 pa.

Canadian Cataloguing in Publication Data

Swift, Jamie, 1951-
 The big nickel

ISBN 0-919946-06-2 bd.   ISBN 0-919946-05-4 pa.

1. International Nickel Company. 2. International business enterprises. 3. Nickel industry — Canada. I. Development Education Centre. II. Title.

HD9539.N52C28   338.7'6'69733   C77-001683-9

Table of Contents

Acknowledgements   7
Foreword   9
Introduction   13
The Birth of a Monopoly   17
Who Dug the Mines?   33
Inco in Guatemala...   63
... And in Indonesia   81
New Growth Strategies: The End of Motherhood   101
Scorched Earth...   113
... And Broken Bodies   123
Why Inco?   143
Appendix: Working in Thompson   153
Afterword   163
Notes   167

Acknowledgements

A number of people have contributed to the research which enabled this study to be completed. Friends at the Latin American Working Group helped with information on the nickel industry in general and Falconbridge in particular. Anne Morrison and the Sudbury Labour History Project lent valuable research information and original interviews. John Lang provided his work on the history of Local 598 and helped in useful discussions of this subject. John Gagnon gave not only the interview but also brought home the urgency of the struggle in a period when there was a risk of being swamped with mere data. Carmel Budiarjo furnished a first-hand perspective on Indonesia. Numerous Guatemalans, both living and dead, uncovered the facts behind Eximbal and Fred Goff of the North American Congress on Latin America helped put them in perspective. Lenny Siegel of the Pacific Studies Centre forwarded important documentation on Indonesia. Floyd Laughren and John Rodriguez, parliamentary representatives for the Sudbury area, provided key documentation. Claire MacKay of the United Steelworkers was helpful with information. Colin Lambert, also of the Steelworkers, contributed material on health and safety.

The original project was devised by the Development Education Centre and the research and writing were done by Jamie Swift. Anita Shilton Martin and Ashley Chester provided valuable editorial assistance and other members of the collective worked on further editing, criticism and improvements. People at Dumont Press Graphix worked on production and Oxfam gave funding support to the project. The project as a whole was a collective one and any shortcomings, omissions or inaccuracies are our own.

Development Education Centre, October 1977

Foreword

by Dave Patterson, President, Local 6500, United Steelworkers of America

As this book goes to print, the Sudbury Basin faces one of the most disastrous blows in its history. Inco has announced its intention of laying off 1800 hourly-rated workers and 400 salaried staff in Sudbury. This is in addition to eliminating the jobs of another 600 people through attrition. As if this wasn't enough, there was the previous announcement of the layoff of over 400 Port Colborne refinery workers. Thompson, Manitoba was also hit with 650 layoffs. These cutbacks have more or less devastating effects on the economic bases of the areas concerned.

Inco's callousness and disregard for those who produce its wealth indicates to me that its responsibility ends at the bottom line of the corporate balance sheet. When a company chooses to cut its production levels by 15 percent the most obvious victims are those who are most expendable — its workers — those whose productivity has put Inco's offices on the fifty-second floor of the Toronto Dominion Centre, in the heart of Toronto's corporate world. Inco workers are now faced with the prospect of further cutbacks as the Company extends its arms around the world to Guatemala, Indonesia and even the bottom of the sea. Inco and other corporations have shown no responsibility for those workers affected by such moves.

Inco is just another four-letter word. This is demonstrated by the obscenities Canadian workers have to endure while governments in Ottawa and the provinces sit on their hands and allow Inco and others to continue their rape of our natural and human resources. This situation can only be corrected when we begin to realize that these resources have to be developed in balance with the welfare of those who produce corporate profits.

This text is an excellent exposé of the corporation known to most of us as Inco. The history of how the Company was founded and how it developed into the powerful organization it is today is almost beyond belief. After reading about how Inco was started and how it prospered, I was certainly given more insight into the complexity of the whole thing. When I read this book, I say to myself, "Not much wonder this Company is comprised of such bastards."

The organization of the Company, from the corporate heads on down, does nothing more than make me realize that I'm a part of this conglomerate. But I belong to the cleanest part. The workers, past, present, and future, have carried the burden of this story around in their heads and have suffered wounds in their hearts and scars on their bodies, hoping that some day Inco would be exposed. The atrocities that have been allowed to take place on the job and in the homes of so many people can finally be printed for a lot more people to actually experience themselves.

I have been a very fortunate individual because I have had the opportunity to work with, drink with, and talk to those individuals who worked for the company prior to the organization of the union. I've heard about bosses firing people for no reason, about payoffs to bosses so that men could keep their jobs (and I don't necessarily mean financial payoffs), about the days when men would stand in line waiting to see if there was any work for the day. I've listened to stories of Company goons smashing windows in the union hall and of Company police beating up workers who stood up to the Company.

Men like Bob Carlin, beat up by scabs because he wanted a union. Union men forced to meet in private, secretly trying to organize the union.

The ruthlessness of Inco goes as far as trying to convince workers that a company union (the Nickel Rash) would solve all their problems. It further demonstrates how far behind Inco was and still is with respect to labour relations. They can hire all the industrial relations whiz kids they want out of the university. It still won't convince me that they're good guys. In the back of my mind I'm convinced there is still them and us. I thank all those older guys and union men who told me all those stories. I certainly learned fast to appreciate and benefit from their experiences with Inco from the pre-union days to the present. I would also like to thank the labour movement for being the best thing this society has for eliminating injustice, both on the job and off.

Things might be a little more subtle now, but the Company is basically still motivated by two things. One is the exploitation of natural resources. Secondly, and most importantly, are the profits which are generated by the extraction of these resources. This must be done free, or relatively free, from government intervention. In the squeeze plays Inco's loyalties boil down to no community, country or group of workers anywhere in the world. Turn a profit large enough to satisfy shareholders, regardless of the cost to workers or their communities. If the government of any one country doesn't fit the bill, change it according to Inco's specifications. If money is needed, put another bank president on the Board of Directors. Or borrow it from the Export Development Corporation.

Although this book goes into great depth in its analysis of Inco as a corporation, and covers the basic story of the drive to organize the corporation, it does not adequately explain the struggles of workers after the famous split and the subsequent merger between Mine-Mill and Steel. Your interpretation of these events depends on your allegiance to either side. It was either a raid or a liberation.

This chapter in Canadian labour history involves a struggle of brother against brother, father against son, family against family. The church was even involved. Inco and Falconbridge just sat back and enjoyed the proceedings. It must have been the most painful experience of any trade unionist's life to have lived through that ordeal. I understand a little about the bitterness because the first man I ever worked with at Inco was Archie Campbell. He was a Mine-Mill supporter and hated the Steelworkers so badly that he refused to attend his daughter's wedding reception because it was held in the Steelworkers' Hall. That same man was the first to put his name on my petition to become a steward for Local 6500. In spite of all the hostility, I believe he eventually did sign a Steelworkers' card. Did we use to have some arguments in 14.70 slusher trench on the 1400 level at Stobie Mine on the 12 to 8 shift! Any free time was spent talking union.

So, eventually Local 6500 of the United Steelworkers of America became the bargaining agent for Inco workers and Local 598 of Mine-Mill stayed on at Falconbridge. Both unions developed as separate entities. As I said, the main problem that I have with the book is that it doesn't explain that the struggle continued after the split for both Locals.

This struggle will always be with us as one between workers and their bosses — in this case, Inco. The reader who finishes this book should go on to explore the links between companies like Inco and the labour movement. If they do not, he or she is missing the intent of the whole thing. Behind every corporation there is the struggle by workers whose story is either waiting to be read or written. Before you can reach a conclusion about any corporation, you must analyze all the facts. It is then inescapable not to conclude that the Incos of the world are taking advantage of a system which allows them to get away with anything they want until such time as the people say, "No more."

— D.P., Sudbury, November 1, 1977.

Introduction

The decades following World War II have been marked by a dramatic increase in the scope and power of transnational corporations — an increase unprecedented in historical terms. In the vanguard of these corporations have been those based in the United States. In 1950 direct investment abroad by American corporations was less than $12 billion. By 1973 the total was over $100 billion.[1] Much of this capital was invested in Canada and in the development of Canada's resource industries. Other funds were funnelled into the underdeveloped nations of the Third World.

Today the assets of these transnationals, firms which carry out production, distribution, research, and sales on a global scale, are well over $200 billion. Yet big firms operating on an international basis are nothing new. In earlier days large merchant companies such as the Hudson Bay Company, the Dutch East India Company, and the Royal African Company were responsible for gathering wealth from faraway lands — wealth which was so important in accumulating the capital which fueled the development of the Industrial Revolution in Europe.[2] Once this relationship between centre and periphery was established and industrial production concentrated in the hands of single-purpose firms, another refinement in the productive system occurred. The small companies of nineteenth century Europe and America were consolidated into large, national corporations as a wave of centralization swept the capitalist world. By the early twentieth century large, hierarchically organized corporations came to dominate production. Nowhere was this process more pronounced than in the United States, where national corporations rapidly expanded their horizons to become transnational corporations. In short—

The development of business enterprise can be viewed as a process of centralizing and perfecting the process of capital accumulation.[3]

However, the reasons for examining the operations of corporations like Inco lie not simply in the size of their holdings or the machinations and intrigues which characterize their operations. Transnational corporations play a strategic role in an international system of inequality and injustice. As such they are specific instruments of oppression. Their effects on people's daily lives are as important as their linkages to other centres of power, the sources of their power and their ongoing strategies. So we have included separate sections on occupational health and safety, labour history, and pollution as integral to the larger study of Inco.

It is part of the motivation behind corporations that health and safety is neglected, that the environment is continuously abused, that people's real needs are ignored. For, as will be shown over and over again, the moving force for business is profit. It is not that those who run the Incos of the world are nasty, villainous ogres who consciously ignore human needs. They, like the victims of the process, are caught up in a system based on profit and growth.

Business is an ordered system which selects and rewards according to well understood criteria. The guiding principle is to get as near as possible to the top inside a corporation which is as near as possible to the top among corporations. Hence the need for maximum profits. Hence the need to devote profits once acquired to enhancing financial strength and speeding up growth. These things become the subjective aims and values of the business world because they are the objective requirements of the system. The character of the system determines the psychology of its members, not vice versa.[4]

Yet there have always been and there will no doubt continue to be those who are in opposition to such a system. In Canada, unions have run into the combined power of the corporation and the state in their attempts to organize workers for better conditions. But their accomplishments have been considerable. Others, both inside and outside political parties, have supported them as part of a broader struggle for meaningful social change. In the Third World, where the most severe human and political costs of corporate growth are borne, struggles against repressive, corporate-oriented regimes continue.

The corrosive effects of the concentrated economic power of corporations on democratic life must be ended. We hope that this study will provide a popular and critical alternative to the public relations image which corporate Canada likes to project. In this way we hope to strengthen the hand of those who are presently challenging Inco and the corporate system it represents.

We realize that many parts of this study are controversial and are likely to elicit a wide range of disagreements. This is perhaps most true of the section on labour history. It is based on an examination of various sources (none of which are "objective") and the evidence of those who have been involved in the struggles at Sudbury over the years. The analysis put forward is an honest judgement based on the facts as we see them. Constructive criticisms are , of course, more than welcome.

Finally, the study is confined in its analysis of Inco in Canada mainly to the Sudbury basin. The stories of Port Colborne and Thompson, as well as those of people who work for Inco in so-called "white collar" jobs, have not been included due to restrictions of time and available energy. Similarly, Inco's operations in other parts of the world, except for Indonesia and Guatemala, have not been researched. We hope people will understand the limitations of such a study and such omissions.

The Birth of a Monopoly

The discovery of nickel deposits in the Sudbury Basin of Ontario in 1856 went unnoticed by all but the most intrepid readers of the Geological Survey of Canada Reports. Not even the government geologist who first verified the find saw any cause for celebration. Yet the re-discovery of these deposits less than thirty years later provoked an entirely different reaction — one that ultimately led to the formation of the International Nickel Company. Today, Inco is Canada's largest mining company and the world's largest producer of nickel.*

For centuries, a nickel alloy called "packtong" was fabricated by the Chinese for decorative purposes. In 1824, its European production was begun under the name "German silver", and after 1860 it was used increasingly for token coinage. With the introduction of silver plating into England in 1884, the use of nickel as a base further increased overall demand. But all of these uses together were not sufficient to encourage large scale nickel production.

The durability and heat resistant properties of nickel had encouraged experiments with it in iron alloys as early as 1822. These increased in the latter half of the nineteenth century and some commercial production had begun by 1885. More important, experimentation with nickel-steel alloys accelerated in the 1880's in France, England and also in North America.

The re-discovery of the rich nickel deposits of Canada occurred in 1883, as work crews blasted a route through Northern Ontario for the Canadian Pacific Railway. This coincidence was precisely the kind of opportunity needed by the businessmen-politicians of the new Cana-

*The International Nickel Company of Canada, long known as Inco, formally changed its name to Inco in 1976. Generally, the two, International Nickel and Inco, will be used interchangeably throughout this book in references to the Company prior to the date of the name change.

dian state to create markets for their modern, bulk transportation system.

It was one of the lesser-known entrepreneurs of the day, an American named Samuel Ritchie, who first perceived the full importance of the Sudbury ore body. He saw a mining industry as a valuable customer for the railway he had been developing in the area. But probably just as important, his past involvement in experiments gave him something more than a premonition of the then unrealized industrial potential of the metal.

> *The introduction of nickel-steel into armaments was the most important single factor in the development of the nickel industry.*
> — J.F. Thompson, former president of International Nickel[1]

In 1886, Ritchie and a group of Ohio capitalists formed a corporation in Cleveland, the Canadian Copper Company, to exploit the nickel-copper deposits of the Sudbury region.

NICKEL COMES OF AGE

A complex of factors — technological, economic, and political — finally gave rise to the production of nickel on a large scale.

The development of steel had been one of the important breakthroughs of the Industrial Revolution, spurring the evolution of rail transport and sophisticated manufacturing machinery. In turn, the efficient use of these new technologies demanded such a scale that the dominant form of economic organization, the small privately owned business, had to be replaced by the joint stock company. Increasingly, these latter were financed by the banks and similar institutions capable of mobilizing the immense capital required.

Railways also encouraged a transformation in the secondary manufacturing sector by weakening local and regional markets. For, as a result of this cheaper and more efficient transport, many small businesses found themselves contending with distant competitors. To survive, locally-based ventures had to pursue the economies of scale and learn to compete themselves in larger regional and national markets.

This mix of technological and business imperatives could not be contained within the boundaries of nation-states. One wave of change fed on the preceding and spawned a third, and unplanned and largely unforeseen, they rippled across the fabric of Western societies. Internal national needs were soon manifested as imperial aspirations. Britain, France, the Benelux nations, Germany, and the United States anxiously vied for both the raw materials essential for their growing manufacturing plants and the markets necessary to dispose of their surplus production. In 1885, this expansionism appeared to reach its peak with the division of Africa by the industrial powers of Europe at the Berlin Conference.

But the appetite of these contending giants for new colonies proved insatiable and military might remained the ultimate means to satisfy it. The modern arms race began in earnest, using to full advantage new technologies. In the process, the centralization of industry was accelerated, since systematic munitions and battleship production required large scale organization. This centralization promoted the further concentration of capital, for only the banks and related institutions could finance such massive undertakings.

Coming full circle, the militarism generated by industrial competition provided the incentive for technological innovation. For example, the French development of armour-piercing chromium-steel shells intensified the long-standing interest in the use of nickel to develop hardened steel armour plate. At the same time, the spoils of militarism enhanced the ability of the colonial powers to protect and expand their empires. France had acquired the South Pacific island of New Caledonia (about a thousand miles off the north east coast of Australia), a rich source of lateritic nickel ore. By 1886, this island had made France, to be exact the Rothschild's Société Le Nickel, the principal nickel producer in the world.

In North America, Samuel Ritchie was embarking on his own strategy for nickel. His Canadian Copper Company acquired the patents to much of the Sudbury deposit. He successfully courted both provincial and federal politicians, as symbolized by the visit of Sir John A. MacDonald to the mine site. In 1889, a 30 percent duty on mining and smelting machinery and a 75 cent per ton duty on smelting coke were lifted by the federal government.

In that same year, a scientific paper on "Alloys of Nickel and Steel" was published in England, clearly indicating nickel's military potential. Ritchie helped publicize this research, bringing it to the attention of the United States Secretary of the Navy, B.F. Tracy. Himself a military man, Tracy recognized the importance of nickel and quickly secured a one million dollar appropriation from Congress to purchase some for experimental purposes.

But Canadian Copper lacked American refining facilities. Indeed it had not yet perfected a refining process. Secretary Tracy solved this problem by turning to Col. R.M. Thompson, a friend who owned a copper refinery in New Jersey. This combination, Thompson's Orford Copper Company and the producing Canadian Copper Company, was the basis for what later became the International Nickel Company.

The connections between these "captains of industry", Ritchie and Thompson, and the U.S. government defence establishment made it possible for their combination to consolidate its control of the American market. Not only did the U.S. Navy provide the main market for nickel; tariff policies were also introduced to favour the consortium. Nickel ore and partially refined nickel "matte" were admitted to the U.S. duty free, while refined nickel carried a ten cent per pound charge.

With its secure U.S. Navy contract and this tariff protection from any European incursions into North America, the Orford-Canadian Copper group would soon be ready to embark on its own expansion abroad. But the divergence between the interests of the mining companies and of the Canadian nation was also becoming clear.

The American duty on the import of refined nickel seriously reduced the prospects for the construction of refining and fabricating facilities in Canada. A pattern was being established, often to be repeated. Resources were being extracted from Canada on terms essentially in accordance with the economic, political, and strategic demands of the United States. The enforcement of these terms was largely dependent on the nature of the corporations involved — put simply, they were American companies ultimately integrated in the American system.

As Americans, naturally we wish to see the refining done in the United States and the work provided for American citizens.
— Stevenson Burke, President of Canadian Copper, 1897[2]

It is ironic that a dispute within the Canadian Copper Company and the ousting of Samuel Ritchie from its board of directors in 1891, gave focus to the Canadian nationalist response to this emerging pattern of development. The American Ritchie, with the support of some Quebec and Hamilton businessmen, suddenly became a strong advocate of a Canadian refining capacity.

His "nationalist" campaign persisted throughout the Nineties and in 1897 the Canadian Parliament even enacted an export tax on nickel ore and matte to offset the American import duty on refined nickel. However, threats from the American combine to move its operations to New Caledonia seemed sufficient to deter the Canadian government from ever proclaiming the enabling legislation. This backdown too, has long since become a familiar pattern in the relations between American corporations and Canadian governments.

In his scholarly analysis of the nickel mining industry in Canada, O.W. Main concluded that "the inability of nationalist sentiments to find effective instruments of control for export industries... strengthened the development of industries in the industrial nation at the expense of the fringe areas."[3]

Despite the sniping by erstwhile competitors, Col. Thompson of Orford felt the combination's position secure enough to permit a serious invasion of the Rothschilds' European market. The development by the German Krupp steelworks of a nickel-steel alloy suitable for armour plate, and the acceleration of the arms race, had made Europe a particularly attractive and important market.

But Le Nickel was not prepared to surrender its European monopoly at the first sign of competition. The result was a price war that lasted from 1892 to 1895, finally ending in an agreement only after Orford had clearly established its superiority. According to some estimates, the North American upstart cornered as much as 70 percent of the world market in 1894.[4] In the course of this extreme competition, several smaller companies attempting to develop other Sudbury-area deposits were destroyed.

THE FORMATION OF INCO

In 1901, J. Pierpont Morgan, using the leverage attendant with his powerful New York investment bank, succeeded in unifying the various components of the American steel industry. Coal and iron mining, smelting and refining operations, wire, plate, and rail manufacture were all integrated into one single corporation — United States Steel.

The constituent parts of this new giant, particularly those involved in armour plate production, were the principal users of the nickel being produced by Orford and Canadian Copper. But this armament production was highly profitable and therefore vulnerable to the entry of competitors. The business logic of this situation did not escape Morgan and he immediately sought to gain control of the supply of nickel needed for the nickel-steel market when presented with the opportunity.

The result was the International Nickel Company, incorporated April, 1902, in New Jersey. For a price of $10 million, Morgan and the financial interests behind U.S. Steel acquired Canadian Copper, Orford, the other major American refinery, the Joseph Wharton Company, and a number of non-operating companies with mining rights in the two largest nickel producing regions of the world, the Sudbury Basin and New Caledonia.

The International Nickel Company, which was organized recently in New Jersey, is the result of plans to consolidate and control the nickel production of the world.
— the *Canadian Mining Review*: 1902

One of the key instruments in this classic concentration of economic power was the Wall Street law firm of Sullivan and Cromwell. Specializing in corporate law, this firm had already assisted Morgan in the organization of U.S. Steel, and the Rockefellers in the consolidation of the Standard Oil trust. Even to the present, a senior partner from Sullivan and Cromwell has always been on Inco's board of directors.

O.W. Main summed up the net effect of the merger: "It provided an assured source of supply for a strategic material and it placed the American armour plate manufacturers in a strategic position to gain a share of the nickel-steel market outside the U.S. while reserving the armour-plate business inside the U.S. for themselves. Also, the strategic control over financing held by the Morgan interests would enable them to block effectively any attempts to finance possible competitors in either nickel-steel or in nickel. Finally, the profits which would be gained from the promotion of the new company promised to be substantial."[5]

THE PERIOD OF MONOPOLY CONTROL

Inco was established on the initiative of the major American consumer of nickel — the steel trust — to ensure supply and hence, stable profitability. In Europe, the other major producer, Le Nickel, was subject to the same kind of pressure, though not so directly, from the Steel Manufacturers Syndicate. With both major producers subject to the influence of their largest customers, there was a reduction in the risk of new entries into the field.* Competition was further reduced by a continuation of the earlier agreement reached between Orford Copper and Le Nickel.

Although the corporations involved often denied it, a cartel was clearly in effect. Even the *Canadian Mining Review*, semi-official organ of the Canadian Mining Institute, stated "that a community-of-interest plan has been arranged which will regulate production, prices, and a division of the markets."[6]

As Inco strengthened its position in the market, Sudbury supplanted New Caledonia as the world's main source of nickel.

	MINE PRODUCTION (Nickel Content — Tons)[7]	
	Sudbury	New Caledonia
1902	5,945	7,000
1907	10,602	6,500
1912	22,421	6,800

*In technical economic terms, the control of a market by the consumer is a monopsony, not a monopoly.

The only successful new entrant into nickel mining in North America was the British, Mond Company, which went into production in the Sudbury area in 1904. Its quick admission to the cartel was based on a number of transparent economic and political considerations. Mond showed no interest in selling in the North American market; nor was it concerned with starting a refinery in Canada, which would have been an embarrassment to International Nickel. Further, it was a family firm, privately financed, with only a secondary interest in nickel production and it did not threaten to have a major impact on total world supply. It readily agreed to market its production through the same agent as the other producers. In short, Mond was the perfect "proof" needed to refute the not infrequent rhetorical attacks in Canada on the "foreign monopolist", International Nickel.

The other aspirants to nickel production in Canada did not fare so well. With the virtual parent-child relationship existing between U.S. Steel and Inco, it was difficult for the latter's competitors to gain access to the primary American market. Equally important was the ability of J. P. Morgan and Company to exercise its considerable influence on Wall Street to forestall the credit necessary to mount a viable large-scale operation. In 1904, the Consolidated Lake Superior Corporation was forced into bankruptcy after failing to raise capital in the New York money market.

The Nickel-Copper Company of Ontario also had every obvious requisite for nickel production — access to adequate ore reserves in the Sudbury region and a commercially viable refining process. Yet, it too failed for lack of financing and markets. Its assets were taken over by the Dominion Nickel-Copper Company, a purely speculative venture by two Canadians, J.R. Booth and M.J. O'Brien. In 1912, these latter sold out to a syndicate headed by a New York promoter, Dr. H.F. Pearson, who had been involved in the financing of Brazilian Traction (now known as Brascan). The new company, the British American Nickel Corporation, included among its backers the promoters of the Canadian Northern Railway, William Mackenzie and Donald Mann.

British America faced the now familiar problems of financing the large scale venture envisioned by its directors and it was not until 1920 that it was able to begin production. This "start-up" was based in part on a British government War contract which had already been revoked by that time. Inco's response to this threat was a price war, which exacerbated the effect of prevailing poor post-War market conditions on the new firm. But it was a law suit initiated by former British America backer, M.J. O'Brien, which ensured the fledgling competitor's collapse in 1924.

Attempts to dispose of the company's assets foundered for almost a year. In 1925, the Anglo-Canadian Mining Company paid $5 million for the physical properties and rights which had been acquired originally with $15 million in cash and $20 million in securities. Anglo-Canadian, of course, turned out to be a dummy corporation set up by Inco. The benefits to the latter were considerable, not only through the maintenance of market control but also in technical gains. For the Hybinette refining process, the rights to which had been held by British America, was more economical than the one originally developed by Orford Copper.

Finally in 1928, International Nickel acquired even the Mond Company. With each company owning rights to part of the extensive Frood deposits in Sudbury, the merger would make extraction of this ore cheaper and more efficient. Mond had also been able to cope better than International Nickel with the vagaries of the nickel market following World War I, and it was pursuing a new strategy of market expansion into the United States.

Prior to this formal combination in December, 1928, the Board of Directors of Inco followed the advice of their Sullivan and Cromwell legal counsel, John Foster Dulles, and converted their company to a "Canadian" corporation by the mere exchange of shares between the former New Jersey parent and the Canadian subsidiary. This move was interpreted as a means of avoiding the anti-trust actions in the U.S. which might have been expected to follow the absorption of the Mond Company.

This consolidation (of Inco and Mond) was like the dynastic marriages which in the past so influenced the course of European history.
— J.F. Thompson, former Inco president[8]

By the end of the Twenties, then, International Nickel was in the formidable position of holding 90 percent of the world nickel market.

THE POLITICAL CONTEXT OF INCO'S GROWTH

Inco's development from its formation in 1902 to its near total monopoly of 1930 had not occurred without problems. The same month that it was incorporated in New Jersey, a demand was voiced in the House of Commons that "all nickel mined in Canada be refined in Canada". In Ontario, the Director of the Bureau of Mines recognized that the creation of International Nickel was a "mere change between American companies".[9]

But Samuel Ritchie and his Canadian allies failed to get either the federal or Ontario government to adopt a policy that would favour refining in Canada. As noted, this failure stemmed in part from the

American consortium's threat to move its operations to the other major known source of nickel, New Caledonia. However, there were more private pressures that might well have "encouraged" government inaction.

Correspondence between Colonel Thompson, president of Orford and later of International Nickel, and Liberal Prime Minister Wilfrid Laurier, includes a claim by the former that he paid Laurier various sums of money, specifically $5000 in profit realized on stock bought for the Prime Minister by the Colonel himself.[10] In later years, large blocks of Inco stock were "transferred" to other prominent public officials, notably two Liberal cabinet ministers.[11]

However, public threats to re-locate and bribes to politicians could not contain the hostility toward Inco that surfaced with the onset of World War I in 1914. In 1913, Germany had accounted for 57 percent of Inco's sales outside the United States. Since nickel was an essential war material, there could be no doubt that a Canadian resource had contributed to the arming of the enemy. Furthermore, the Company did not (and could not) deny that shipments of its finished nickel were continuing to go to Germany via the neutral United States, even after the War had started.

The present situation is that Britishers and other enemies of the Teutonic armies are being shot down by machine guns hardened with Ontario nickel and not an Ontario boy goes into action except at the risk of having to face bullets barbed with the nickel of which his own province has a monopoly.
— *Toronto Telegram*, December 27, 1914[12]

In an atmosphere heavy with wartime propaganda, Inco's public image reached its nadir in 1916. Even the Financial Post was motivated to evaluate the Corporation's tax rate, finding it to be a mere 3 percent. This general uproar allowed James Conner, the social democratic candidate for South West Toronto in a 1916 provincial by-election, to extend the criticism to Inco's labour relations, albeit somewhat circuitously: "Never was tyranny so outrageous as that shown by the nickel trust. It is not satisfied with making its easily begotten profit from enemies of this country. It is not satisfied with being blatantly and openly guilty of trading with the enemy, but it even adds to its soulless character in treating its workers with a brutal despotism."[13]

The business press took an equally critical view of International Nickel, though for quite different reasons: "No honest far-seeing investor, or public man will have any good to say of the nickel company and its alleged methods, for the reason that they reflect upon and do serious damage to the reputation of all Canadian corporations. These methods afford the excuse for agitators, demagogues, self-seeking, unscrupulous politicians, and newspapers like the Toronto Telegram to go among the masses... and secure support by condemning all private enterprise from large corporations to small manufacturers."[14]

To the relief of those holding this opinion, the Liberal opposition in Ontario, not Conner and his ilk, won the two legislative seats at stake by basing its campaign, too, at least in part, on the "nickel question".

Later that year, an American newspaper revealed that the submarine *Deutschland* had made two trips, the second in November, 1916, transporting in total 600 tons of nickel from the U.S. to Germany. Realizing its precarious position amid growing demands for the expropriation of its Canadian assets, Inco announced plans to construct a refinery at Port Colborne, Ontario, adjacent to the cheap power supply of Niagara Falls.

Considering that a new refinery in Bayonne, New Jersey had only been opened in 1913, this sudden decision was a clear admission of the value the Company placed on its future in Canada. As such, it exposed the weakness of previous threats to abandon Canada for Inco's "vast" holdings in New Caledonia. But this move did not increase the national control over the ultimate use of Canadian resources, for other refineries were retained outside the country. And the federal government took no action to encourage or coerce the development of a nickel-steel industry in Canada.

In 1917, the Ontario government felt bold enough to introduce a New Mining Tax Act, which added to the longer term incentive to refine in Canada, by calculating mining profits only on the basis of the sales of refined metals. It did not allow the deduction for Ontario tax purposes of taxes paid by the Company's U.S. refineries, nor did it accept as legitimate, sales of ore or matte to "associated companies" in the United States.

Only the War, with its undeniable social pressures and the short-term yet extraordinary economic leverage it conferred on this country as the monopoly source of a strategic metal, had been sufficient to prompt a governmental challenge to the power of International Nickel.

THE CONSEQUENCES OF THE PEACE
The end of World War I brought a general recession with an attendant decline in corporate profits and massive unemployment. And the well-being of the nickel market, even more closely than in other industries, was directly related to military spending. In 1921 and 1922, International Nickel reduced its extraction of ore to levels less than five percent of the tonnage raised in 1918. Only the more stable operations

of the Mond Company prevented the total social disintegration of Sudbury.

This stunning decline resulted in a re-organization of Inco's management and a concerted effort to research and develop peacetime uses of nickel. The automobile, particularly as a mass-produced consumer good, was the most important technological and social innovation of that time; as such, it provided many new opportunities for nickel use, in everything from transmissions to decorative chrome trim. By the mid-Twenties, nickel was firmly established as a consumer-related product, with 36 percent of all its U.S. consumption by the auto industry.[15]

In Germany, Krupp, the massive trust which had been the backbone of the arms industry in that country, developed stainless steel, a new multi-use alloy that contained nickel as a main element. These and other innovations eased nickel's transition from a strategic material in the narrow military sense to broader significance in modern industrial economies.

When the boom of the late 1920's collapsed to the bust of 1929 and the Great Depression, even Inco, with its virtual monopoly in the nickel market, felt the effects. From 1929 to 1932, production dropped drastically, as the Company bore the brunt of the fall in American demand. In Europe its sales plummetted by 75 percent. Its workforce was decreased from 7181 in 1930 to 1490 in 1932.[16]

However, 1932 was the last year International Nickel failed to record a profit. For the war-related demand for nickel was being renewed, both in Europe and in Asia. Nations prepared for the armed conflicts to come — the Italian invasion of Ethiopia, the Italian and German fascist intervention in the Spanish Civil War, the Japanese attack on China.

In 1934, John Foster Dulles, senior partner in the law firm of Sullivan and Cromwell and general counsel for Inco, negotiated an agreement with I.G. Farben, the German trust. Farben was guaranteed up to 10 percent of the total sales of the Inco subsidiary, Mond, in return for which it would buy exclusively from the latter and would not attempt to develop any nickel mines.[17] This allowed Germany, without any significant reserves of its own, to stockpile Canadian nickel for military purposes.

Sullivan and Cromwell, one of America's largest corporate law firms had strong historical ties with German companies and the German government. Its Berlin office brought to the firm a significant amount of business. However, as the persecution of the Jews intensified, the nature of Hitler's fascist regime became unavoidable. By 1935, most of the partners in Sullivan and Cromwell wanted no more dealings with Nazi Germany. When they threatened to resign over the issue, "Foster Dulles was at first bewildered, then adamant, protesting among other things, the loss of substantial profit; finally, however, he capitulated 'in tears'."[18]

In Canada, the sale of nickel abroad at a time when the likelihood of war seemed to be increasing, once again aroused public criticism. In 1935, the Royal Canadian Legion asked for an embargo on nickel sales to European armament makers. In 1937, the Trades and Labour Congress urged the nationalization of the nickel industry.

However, Inco's campaign since World War I to "Canadianize" had apparently worked. The influential *Financial Post* seemed to forget its past concerns with the Company's behaviour; on the contrary, the fiftieth anniversary of the discovery of nickel in Sudbury inspired it to editorialize that it was "... fifty years ago, that the Canadian Pacific Railway, in gouging out its right-of-way, through the Sudbury area, gave birth to the great mining industry we now know as the International Nickel Company Ltd., ... in the past, this farflung company has proven itself to be a benevolent monopoly."[19]

Despite debates in the House of Commons every year from 1934 to 1937, on the desirability of controlling nickel as a strategic war material, the Canadian government remained sympathetic to the "new Inco", accepting the argument that it would be impossible to control the "ultimate" destination of the Company's product. Nickel deliveries direct from Canada to Japan and Germany were, therefore, allowed to continue at their normal level.

> *Assuming that moderate expansion in world industrial activity is to take place in the coming twelve months, the consumption of nickel should attain a volume higher than ever before. Should there be an American boom next year in such industries as automobile, housing, and the "heavy" industries ... the demand for nickel will increase very substantially. There are good possibilities that billions of dollars may be spent by the major powers in increasing their naval and military appropriations.*
>
> *It is clearly evident that International Nickel will profit to the fullest extent from the substantially increased demand which world-wide industrial recovery will create for nickel....*
>
> — Analysis of International Nickel by Financial Services Ltd., Montreal, 1935

THE FALCONBRIDGE CHALLENGE TO INCO

Almost simultaneously with the merger of International Nickel and Mond, an eccentric Canadian mining promoter named Thayer Lindsley had been organizing a new entrant to the nickel business.

Lindsley's venture, Falconbridge, was modest, requiring only $5 million, a sum he was able to secure with the backing of his personal American contacts. It would have been difficult for International Nickel to have blocked the financing of such a small scale project. Though the new company's mines were on the few properties in the Sudbury region not controlled by Inco, its refinery was in Norway, and its search for customers was concentrated in Europe. Thus, Falconbridge escaped the fate of its predecessors, the British America Company, the Lake Superior Corporation, and the Nickel-Copper Company. For not only did it not pose a direct threat to Inco's North American monopoly, any attempt to eliminate it would have invited anti-trust actions against International Nickel, its subsidiaries and customers in the United States. This danger increased in 1932, after the election to the presidency of the reformist Franklin Delano Roosevelt.

Falconbridge enjoyed a slow but steady growth, reaching a peak profitability in 1939 of $2 million on sales of $8.2 millions as it supplied nickel to a war-ready Europe. Ironically, the actual outbreak of war curtailed Falconbridge's profitable operations — in 1940 the German Army occupied Norway, appropriating the company's refinery at Kristianssands.

Falconbridge's ill luck was Inco's good fortune, the latter agreeing to refine the former's entire ore output and doubling its own ore tonnage, all in the name of the "War Effort".*

The first obligation of every corporation, as of every individual, is to give the utmost support to his (sic) government in the prosecution of the war. . . . In the vigorous performance of this duty, we shall continue to develop the full measure of our resources and experience.
— Robert C. Stanley, president of Inco[20]

The end of the War brought challenges to International Nickel which favoured the growth of Falconbridge. In 1946, the U.S. Department of Justice filed suit against Inco's American subsidiaries, charging monopolistic practices. This matter was settled in 1948 with a consent decree which allowed the Company to maintain its own rolling mills as long as it supplied other mills at competitive prices.

An apparently more serious threat came from the American Defence Department. The demand for nickel had increased dramatically

*Wartime controls and taxes restricted increases in corporate earnings. Nonetheless, Inco's assets increased dramatically as sales and production rose. With the fall of France, New Caledonian matte was also diverted to Inco's refineries, making it the totally unchallenged supplier of nickel to the Allies.

during World War II, prompting the designation of the nickel industry as one of "mandatory priority status". This gave to government control of the allocation of the "war metal", as it came to be called. Following the War, U.S. government planners forecasted a long term and significant growth in nickel demand, not only for military purposes, but also for industrial uses. Thus, the American government, a mainstay of the Company's prosperity since the first days of the armour-plating experiments, decided to encourage other sources of production.

A key element in the diversification plan was an agreement by the American government to buy 100 million pounds of nickel from Falconbridge at a price 40 cents per pound above the prevailing market rate, a clear subsidy of $40 million. With this guarantee, it became easier for Falconbridge to procure the private financing for the expansion which was to establish it as the second largest nickel producer in the non-Communist world.

Other corporate beneficiaries of this American stockpiling were the Sherritt-Gordon Company of Canada and the Freeport Sulphur Company of the U.S. The former was able to develop operations in Manitoba and Saskatchewan. Freeport Sulphur had the development of its Cuban holdings underwritten by the U.S. government, though these were nationalized in 1960 by the revolutionary government led by Fidel Castro.

Between 1950 and 1957, the United States spent $789 million to stockpile nickel, in the process eroding Inco's previously overwhelming monopoly position. However, the primary intention of this policy was not to attack the nickel giant; rather, it was to stimulate production in order to meet the long term military and economic needs of the rising American empire.

The end of the war had not brought the massive reduction of military expenditures which could have been expected. The dominant U.S. policy planners understood the opportunities for their nation to expand its economic influence with the collapse of all its former competitors in Europe and Asia. They were not slow to recognize the military demands such expansion would impose, either. Conveniently, the early manoeuvering in their campaign rapidly escalated into the Cold War, providing a strong ideological cover, anti-Communism, for military build-up throughout the Fifties. The Korean War gave further reality to what had at first appeared to be a vague contest of ideologies. And the refinement of jet aircraft and nuclear weapons, two new heavy consumers of nickel first introduced to the modern military arsenal during World War II, continued apace.

Just as important in U.S. policy making was the anticipated growth in nickel demand for non-military uses. Normal economic mechanisms could not be depended upon to meet long term demand, which was expected to grow at an annual rate of 6 percent. Incentives to the one major producer (in this case Inco), would not only have been impolitic, but also unreliable. For the long term interest of a monopolist is surely the maintenance of a "tight" market in which demand in excess of supply justifies premium prices.

Consequently, the internal health of the American economy as well as the external opportunities for its expansion, required this massive subsidization of International Nickel's competitors. And the record of the Company's profitability over the past three decades demonstrates that that general prescription has not been incompatible with Inco's specific well-being.

Who Dug the Mines?

Even as J. Pierpont Morgan was organizing the merger of Canadian Copper and the Orford Company from his Wall Street office, a few of the Sudbury mine workers were attempting some organizing of their own. But the outcome of their efforts was negligible, paralleling the fate of many similar attempts to be made across Canada in the next four decades.

Unions, particularly those which attempted to organize locals in the primary production sectors of the economy, faced an overwhelming combination of forces in opposition to them. The very conditions that demanded collective organization — the twelve hour day, hazardous work conditions, and frequent layoffs — sapped the energy needed by the workforce to organize itself. Most companies, Inco included, also did everything possible to thwart unionization.

> *The International Nickel Company had hired stooges from the Pinkerton Spy Agency, and they got jobs in the mines and the smelter and would encourage union talk or try to get you to talk about the union. And as soon as you talked to them, you were signing your warrant (to be fired).*
>
> — a Sudbury miner[1]

In their efforts, companies usually had the willing assistance of the various levels of government, the judiciary, and the media. For more direct intimidation, they hired goons and spies, particularly through the notorious American-based Pinkerton Agency.

The nature of the Sudbury workforce, indeed, of most of the North American industrial working class, provided ample opportunities for the use by company management of "divide and conquer" techniques. The hinterlands of Canada were largely populated by an heterogeneous mix of immigrants. In the case of Sudbury, most management and some skilled workers were of British origin, while most of the workforce consisted of more recent immigrants, Poles, Finns, Ukrainians, Swedes, and Italians. This immigration had been complemented by internal population shifts, Francophones from both Quebec and other parts of Canada adding an important element to the resulting linguistic, religious, and cultural mix.

Following the pattern already established elsewhere in North America, Inco capitalized on the tendency of immigrants to sustain, almost to the point of isolation, their own separate communities. This use of hiring policy to encourage insecurities based on ethnic differences, a practice intended to militate against "on-the-job" solidarity, was called the "Rockefeller formula" in labour circles.*[2]

To further complicate the situation, some ethnic communities harboured their own bitter, internal divisions, "Red" Finns against "White" Finns, "White" Ukrainians against "Reds", Irish Protestants against Catholics.

> *I had grown up with Inco telling everybody what to do. It just got under my craw to think that if I wanted to go out and play with certain kids, well my dad would get the blast — "What's your daughter doing down there, playing with those Finn kids or those Polish kids?"*
>
> — Peggy Racicot[3]

Another factor which hindered drives to unionize was the instability of employment. Prior to the massive state intervention in capitalist economies that has been called the Keynesian revolution,† the business cycle tended to the extremities of boom and bust. The primary industries, including the resource extractive ones, were especially vulnerable. A standing joke in Sudbury was that there were three shifts at Inco — one getting hired, one working, and one getting fired.

THE STRUGGLE TO ORGANIZE

The main union of hardrock miners in North America at the turn of the century was the Western Federation of Miners, which had been

---

*It was an ambitious young Canadian politician, W.L. Mackenzie King, who spent five years (1914-1919) helping John D. Rockefeller Jr. systematize a "labour strategy", in the process winning the American tycoon's lifelong friendship.

†Though Keynes may have instructed state planners and economists in the importance of using government fiscal policy to reduce the extreme effects of the business cycle, he provided no long term solutions, as should be clear from the present state of Western economies and Keynesian economics.

founded in Butte, Montana in 1893. One of the first known attempts to organize in the Sudbury Basin occurred in 1905 when a local of the WFM was formed. Though that local did not survive, "the Western Fed" became popular in other parts of Northern Ontario. In Cobalt, mine workers succeeded in establishing Local 146, a success which encouraged Sudbury-area workers to persist in their efforts. During this period, the WFM granted charters to the Garson Miners' Union as Local 182 and the Coniston Smelterworkers' Union as Local 183. Yet neither was strong enough to withstand the attacks of Inco's management and according to one activist they never became more than "paper locals".[4]

Around the time of these early efforts at unionization in Sudbury, the Western Federation of Miners in the United States was leading an attempt to advance the goals and methods of the labour movement beyond its traditionally defensive posture. With 27,000 members, the WFM provided the largest base for the formation of the Industrial Workers of the World. From the miners' ranks also came some of the IWW's most effective leaders, including the now legendary "Big Bill" Haywood.

The first constitution of the IWW began with a ringing statement of its basic position. "The working class and the employing class have nothing in common . . ." It also stated this new organization's strategy with admirable clarity: "Between these two classes a struggle must go on until all the toilers come together on the political, as well as the industrial field, and take and hold that which they produce by their labour through an economic organization of the working class, without affiliation with any political party."

The fortunes of the IWW and the WFM fluctuated in the decade preceding World War I, both in Northern Ontario and throughout North America. In 1908, the WFM severed its ties with the IWW, though Bill Haywood remained with the latter. The IWW went on to lead some of the most important and dramatic campaigns in the U.S. prior to the Great War, though it was unable to maintain an organized base. The Miners' union suffered a period of decline apparently related to a drift to conservativism which was symbolized by its re-affiliation to the AFL in 1911.

Perhaps the violent opposition encountered by the IWW, as well as its successes, reflected the perception of societies as "class societies" then more common than in our own time. For even J.P. Morgan, the financier who had organized the Inco monopoly, held the view that classes existed in opposition to each other. In 1908 he called a recent depression " . . . a good thing. A good thing to prove to the working classes that they cannot control wages, this being the exclusive prerogative of the employing class."[5]

The American entry into World War I signalled an all out attack there on the IWW as a subversive organization. And Canada's earlier inclusion in this War (as part of the British Empire), had already provided companies like Inco with the opportunity to openly use racism in their campaigns against unionization.

In the course of one such attack against a drive led by workers of Finnish origin, the *Sudbury Star*, consistently faithful to the interests of the mineowners, sought to capitalize on the climate which had been developed by so much of the War propaganda. It described the arrest of one "smooth young foreigner" who had been addressing a rally of 67 other "foreigners".[6] Most, if not all these "foreigners" were Canadian citizens, though that did not prevent a War-time prohibition on gatherings of their organization and against publication of their Finnish-language literature.*

But International Nickel did not rely only on the extraordinary methods of repression sanctioned by World War. Infiltration of the workforce by "Pinkertons" continued. In 1916, two hundred workers were fired, their pay envelopes scrawled with the message — "We don't want troublemakers."[7] As a result of this sort of intimidation, Sudbury soon acquired a reputation in the mining towns of the North as a haven for spies, scabs, and company stooges.

As the one WFM local in Sudbury, the Coniston Smelterworkers' Union was disintegrating in 1916, the Federation was facing declining fortunes continent-wide. Its response was a name change to more accurately reflect the union's jurisdictional aspirations. Thus, the International Union of Mine, Mill and Smelter Workers (IUMMSW or Mine Mill) was born, carrying within it the memories of a militant tradition and re-affirming a commitment to what was then the contentious notion of "industrial unionism".

In the nineteenth century (and earlier), workers had organized on the basis of their craft, reflecting the small scale of most early enterprises. However, as the dominant organization in maturing capitalism became the large scale corporation, organizing many different skills within the same complex, craft unionism tended to divide workers within the same industry rather than uniting them. Further, the craft unions defended their own form and the interests of those skilled craftsmen which they did represent, by vigorously opposing the de-

---

*The irony of this kind of jingoism on behalf of a company that allowed the shipment of some of its production to Germany in the midst of the War, is one of those instructive footnotes to history.

velopment of other forms of unionism. Through the American Federation of Labour and its affiliate in Canada, the Trades and Labour Congress, conservative unionism dominated the entire North American labour movement.

Some form of industrial unionism, that is, the organizing into one union of all workers in one industry, regardless of their particular skills, was the obvious alternative to craft unionism.

Along with its industrial unionism, Mine Mill inherited from the WFM a less easily defined "political tradition". As exemplified by the actions of those mine workers such as Haywood who participated in the IWW, this was syndicalism — an approach based on the theory that all workers should organize into one union to take over the means of production. Syndicalists opposed the formation of political parties, considering them an unnecessary and harmful separation of the immediate problems and conditions of the producers — the workforce — with the goal of a transformed society, self-managed by the collective producers.*

Mine Mill drew from its common origins with the "Wobblies" a more general commitment to a "politicized approach" to the long term interests of its members. This stood in contrast to the policy of the craft unions in the AFL. Under the leadership of Samuel Gompers, they had perfected the art of "getting along" with any government, regardless of its ideology, in order to defend the immediate and short-term interests of their members.†

THE BIRTH OF MINE MILL LOCAL 598
In 1919, at the end of the War-induced boom in the nickel industry, an attempt to organize in Sudbury led to the formation of Mine Mill Local 116. Within a year, this local had gone the way of its predecessors and within two years the bottom had fallen out of the nickel market, reintroducing the insecurity of the "boom-bust" cycle.

From 1923 to 1928 nickel mine production in Canada recovered, steadily growing to the level of production achieved in 1918. Canadian production finally exceeded the wartime record in 1929; but the effects of the Great Depression heralded by the stock market crash on Black Friday, October 24 that year, were not to by-pass Sudbury.

*Even the IWW was not a consistently syndicalist organization, containing within it divergent perspectives on political parties.
†The weakness and inchoate nature of this more politicized approach of the WFM and its progeny should not be romanticized out of consideration. Mine Mill was plagued by an internal "left-right" split, just as the WFM vacillated between support for radical experiments such as the American Labour Union in 1902 and the IWW in 1905, and affiliation to the AFL in 1911.

The first response of workers across Canada to the Depression was understandably unorganized and confused. It was impossible to foretell what was in store. The Trades and Labour Congress proclaimed its own impotence, insisting that it was impossible to organize the unorganized or even to lead a successful strike in times of depression. To fill the leadership vacuum, the Canadian Communist Party formed its own trade union centre in 1930, called the Workers' Unity League (WUL).

In 1933, the Mine Workers' Industrial Union, an affiliate of the WUL, tried to organize in Sudbury but failed. However, its organizers met greater success across Canada, having already led the Estevan coal miners' strike in 1931 and going on to organize the Flin Flon general strike in 1934.

Though these particular struggles were violently repressed, they combined with many similar drives, both Communist-led and independently organized, in Canada and in the United States, to prepare the path for a second larger wave of unionization in the latter half of that decade of Depression. For this increasing combativeness of the workforce pressured the more astute leaders of the established unions to re-examine their position; it also warned the political guardians of the established order to reconsider their reliance on brute repression to maintain social control.

International political developments added one more positive factor to this favourable mix. Desperately eager for alliances that might enhance the security of the Soviet Union, the Communist International ended its attack on social democratic parties and ordered its member Parties to abandon their independent labour strategy and integrate their efforts into the established labour movements. In 1935 the WUL was liquidated and its affiliates' locals were turned over to TLC unions. But the main diversion of energies prompted by this decision was on the level of individual Communist union organizers. Developments in the American Federation of Labour promised them a somewhat more compatible "home" than conservative craft unionism.

John L. Lewis, the dictatorial leader of the United Mine Workers and a political conservative, had started a campaign within the AFL for an "official" drive along industrial lines. He had been very cautious throughout his campaign, even refusing support to independently organized local unions in the American "steel belt" because he could not control them. But, when the October 1935 convention of the AFL refused to sanction industrial organizing, Lewis, with the support of eight unions including the Amalgamated Clothing Workers, the International Typographical Union, Mine Mill, and his own United Mine Workers, formed the Committee for Industrial Organization. In 1937 this committee became a separate trade union centre, the Congress of Industrial Organizations (CIO), when the AFL expelled those unions that supported it.

It was to these unions and the organizing committees they established, that the experienced organizers of the WUL gravitated.[8] Throughout North America campaigns got under way in the name of the CIO, in the steel, auto, textile and rubber industries, in packing houses, electrical and machine shops, and in mines. Despite the use of the name, in Canada these drives were financially and organizationally independent of the U.S. Committee.

In 1936, Mine Mill Local 239 was chartered in Sudbury. One hundred and fifty mine workers quickly signed cards, a definite gain over previous efforts. George "Scotty" Anderson, a veteran of the WUL, was the only full-time Mine Mill organizer and as such he was responsible for the entire push in Northern Ontario. Though he and his members were able to establish a monthly information sheet, called the *Union News*, International Nickel was quick to defuse this threat to its supremacy.

In 1936-37, the Company conceded wage increases totalling 15 percent. Of course, it coupled the "carrot" with the "stick" — Anderson was attacked by a gang of thugs and the Mine Mill office was broken up. Combined with the union's shortage of funds, Inco's "persuasion" curtailed the growth of Local 239. The international office could not even continue to pay Anderson and the CIO had not yet been established in Canada as a real organization capable of assistance. Without a full-time organizer, membership dwindled in face of the Company's constant intimidation. In one belated effort to recruit at the Creighton mine only six workers were signed. By 1938, the Sudbury local was defunct.

In April, 1937, following the stunning success of a "spontaneously organized" strike in Oshawa against General Motors, the CIO opened a Canadian office and the wave of unionization which had begun in 1935 regained momentum.* Nevertheless, it took a 12 week strike in Kirkland Lake, beginning on November 18, 1941, to renew the impetus for organization in Sudbury.

This strike attracted national attention because of the conduct of the companies involved. Hollinger Mines and the other gold producers of the area refused to recognize the Mine Mill local, though recognition seemed to be required by federal law and had been recommended in

---

*The contradictory term "spontaneously organized" is used only to indicate that that strike was organized without assistance from the outside, that is, without the help of professional organizers.

any case by a federal conciliation board. Instead, the mineowners enlisted the support of the Ontario government, which sent in special police units to "protect the mines". Liberal Premier Mitch Hepburn, a former onion farmer, had previously demonstrated his alarm at the prospect of the CIO entering Northern Ontario.*

> Hepburn shared the same world of robust and roistering stock promoters and mining barons . . . his circle of friends seemed to be drawn mainly from this group of men with particular interests in the Ontario economy.[10]

The Kirkland Lake strike had been called against the advice of the leadership of the Canadian Congress of Labour (CCL), the union central which had been formed in 1940 by the CIO and other industrial unions already existent in Canada. The Toronto-based company owners took every opportunity to stress that while the strikers were mainly "foreigners", those who stayed on the job were of "British or Canadian" origin.[11] However, the blatant alliance of the provincial government and the police with private interests and the refusal of the federal government of Mackenzie King to enforce its own laws, served to win broad public support for the strikers from across the country. The labour movement contributed several hundred thousand dollars to the strike fund and the strike continued on into a bitter Northern Ontario winter. By February, 1942, the local had to admit defeat and its members went back to work. But by then the example of the Kirkland Lake workers was adding to the growing spirit of militancy sweeping the mines of Northern Ontario.

> In Sudbury, many workers unable to afford the cost of transportation to and from the mine heads, were forced to squat in shacks right on Company property. Electricity and plumbing were a rarity, and the women had to carry water from nearby streams.
> Conditions on the job were equally deplorable. The 56 hour work week prevailed, with no premiums for overtime or shift work. Paid holidays, health, welfare, and other fringe benefits did not exist. Accidents were a regular occurrence. The "Levack Tragedy" of 1938 took the lives of five miners when the skip (elevator) they were riding hit a timber. Several others were seriously injured. The wife of one of the casualties recalled: "Our life was just starting and it stopped right there."[12]

---

*During the Oshawa strike, Hepburn seemed more concerned that a "CIO" success would lead to the "infection" of the mineworkers of Northern Ontario, rather than that it would damage General Motors.

In an effort to undercut the appeal of Mine Mill, International Nickel had formed a company union — a time-honoured anti-labour tactic. It began by offering a "contract" to its own Employees Welfare Associations in Sudbury and Port Colborne. Several representatives of the Associations then met with Inco vice-president Donald McAskil in his office. The result was the so-called United Copper-Nickel Workers.

This facade of unionism, very much in the Mackenzie King tradition of "labour representation", soon became known among the Company's workers as "the Nickel Rash" — a minor affliction. In 1942, Inco spent $68,000 trying to spread the "Rash". However, it was mainly able to muster support for its creation by offering promotions to those who joined.[13]

In February, 1942, a delegation of Sudbury workers went to Kirkland Lake to ask Mine Mill organizer Bob Carlin to lead a campaign which had already begun at Inco. In the face of the activity that followed, the Company resorted once again to its less refined "union-busting" tactics.

On February 24, a dozen "goons" visited the new Mine Mill office in Sudbury. Among these visitors were several Inco shift bosses, apparently "working" at the time. They demanded to see Carlin, the principal organizer, but he was home sick. In frustration, they tried to get the names of those who had joined the fledgling local from the two people in the office, Forrest Emerson, another full-time organizer, and Jack Whelahan. Fortunately Carlin, an experienced organizer, had put the local's records in a safety deposit box.

Failing to get satisfaction, Inco's "industrial relations representatives" proceeded to break the office's windows, destroy its furniture, and tear out its phone. They used a typewriter to smash Emerson in the head and body and they methodically beat Whelahan, who continued to resist. One Mine Mill activist who visited the latter in hospital the next day, reported that Whelahan's face resembled "an old rubber boot".[14]

Although the union office was in the busiest part of town, an area regularly patrolled by the police, no servant of the law was around at the time of the attack. It was later ascertained that the police had been ordered off the street by the municipal authorities in anticipation of the event.[15] When questioned in the House of Commons about this incident, the Ministers of Justice and Labour both claimed that law enforcement in Sudbury was the sole responsibility of the local police. The RCMP concurred.[16]

This brutal attack forced Mine Mill to ensure that its activities be as little exposed as possible to Company surveillance and harassment. It distributed ten thousand handbills describing the attack, effectively taking advantage of the sentiment that anything which Inco felt it necessary to crush might well be good for the workers. However, no public meetings were held as the local sought to "disappear".

Instead, the organizers and supporters secretly deposited handbills explaining the union in public places, such as hotel washrooms. If a positive reaction was detected from a patron who had noticed the union's literature, a direct approach was made. Sympathetic workers were methodically organized into small, independent groups, further minimizing the danger of infiltration by Company agents.

These tactics soon bore fruit. The local quickly grew to several hundred members although the Company's influence still made it difficult to book meeting halls. But increasingly, the unionists became strong enough to go on the offensive, in January, 1943 publishing their own weekly newspaper, the *Sudbury Beacon*, in opposition to the *Sudbury Star* (the existing paper which was more popularly known as the *INCO Star*).

Further, the union made an effort to recruit the women who were working at Inco because of the labour shortage being caused by World War II. Many of them were eager to join the union, both because they were being paid less than men for equivalent work and for other less tangible reasons.

The shift boss called me in and told me that there were rumours that they were going around trying to organize a union. He said, "Your father being a shift boss, you just better not join that union." I said, "Mr. Ferguson, are you asking me or are you telling me?" He said, "I'm damn well telling you. Don't let me hear of you having anything to do with them." So I said, "Thank you, Mr. Ferguson," and I walked right out there and walked down the back track and got a hold of (the union organizer) . . . and I said, "I want to join the union" . . . I was eighteen then.
— Peggy Racicot, Inco wartime worker[17]

This changed atmosphere related to nation-wide and world-wide developments during World War II. The social democratic Co-operative Commonwealth Federation (CCF) rode the political crest of this discontent, becoming the official opposition in the Ontario election of 1943 and the following year becoming the government in Saskatchewan.

In Sudbury the relationship between political and union developments seemed to be symbolized by the election of Bob Carlin to the provincial legislature in 1943, with more votes than any other CCF candidate in that election. But Carlin's victory had other significance.

Since 1940, the CCF in Sudbury, led by Jack McVey and James Kidd, had been building an organization, eventually counting 2700 members in their club. On the orders of the provincial leadership, the CCF activists assisted in Mine Mill's organizing drive. As a young mine worker and president of the Sudbury CCF, Kidd seemed to be the logical candidate for his party's provincial nomination in 1943. But after changing his mind three times, Carlin decided to seek the nomination, prevailing upon Kidd to step aside in order to avoid weakening the image of the union. Thus, at a time of triumph on both union and political fronts, the seeds had been sown for a personal rivalry that would have serious repercussions.

By July, 1943, Mine Mill Local 598 was well enough established to host the first labour conference in the history of Sudbury — the Canadian Conference of Metal Miners. From this forum, both Inco and Falconbridge were condemned for placing a higher priority on suppression of unions than on efficient production for the War. For, despite a shortage of experienced mine workers, both nickel producers refused to hire experienced men from other northern centres, for fear these might be union activists.

Disregarding Mine Mill's obvious success, Inco steadfastly maintained that it had a contract with the United Copper-Nickel Workers and refused recognition to Local 598. However, the belated recognition by the Ontario government of the growing strength of the labour movement led to the passage of a Trade Union Act in 1943. And, in a certification vote in December of that year, 6913 Inco workers supported Mine Mill, while 1187 opposed it. At Falconbridge, the tally was similar, 765 to 194. The 1400 International Nickel workers at the Port Colborne refinery were also successfully organized by Mine Mill, 90 percent of them having signed cards for Local 637, by May, 1943.

On March 10, 1944, Inco signed its first union contract. Because of wartime wage controls, this agreement did not provide for a pay hike. However, the Company had been forced to accept a grievance procedure, the eight hour day, seniority rights, paid vacations, and a voluntary check-off of union dues. Local 598, the largest in that entire international union organization, seemed well on its way to stability and Mine Mill to establishing itself as the principal miners' union in Canada.

MINE MILL UNDER ATTACK

The series of attacks which were mounted against Mine Mill, starting in 1949 and continuing for over a decade, must be considered in the context of union politics, the Cold War, and the role of its biggest local, 598, in the community of Sudbury.

As one of the eight international unions which had formed the CIO in the United States, Mine Mill had attracted many skilled and dedicated Communists. Though this same situation occurred in many other CIO unions and organizing committees, in a few unions, including Mine Mill and the United Electrical Workers (UE), the Communists eventually dominated the international leaderships.

In Canada, the Communists faced two other more or less coherent approaches to unionism and the acrimonious conflict that eventually resulted became inextricably interwoven with the politics of the nascent Cold War. The strongest of these two other currents in the CCL was the one that shared with the Communists a conception of unionism which demanded a commitment to political involvement. Its leaders were social democrats, members of the CCF, and it was best typified by the United Steel Workers of America (USWA). Charlie Millard, a veteran of the 1937 strike against General Motors whose first job as Canadian director of the Steel Workers' Organizing Committee was to rid it of Communist Party members, became the main tactician for this tendency in Canadian unionism.

The other current, less frequently articulated, was cautious to the point of conservatism and might be characterized as "pure unionism". In the Canadian Congress of Labour, this tendency was best represented by Pat Conroy, its full-time and powerful secretary-treasurer. Quite simply, he distrusted both "political factions".

The vulnerability of the Communist Party unionists stemmed in part from their total identification of the long term interests of the Canadian working class with the short term needs of the Soviet Union, which they saw as the "motherland of socialism". The best example of the confusion and antagonism that this caused is their attitude to the Second World War.

Because of the "Hitler-Stalin Pact", a non-aggression agreement between the U.S.S.R. and Germany, Communist Parties throughout the world opposed World War II, characterizing it as merely another "inter-imperialist war". On June 22, 1941, the German Armies invaded the Soviet Union and the position of the Communist International instantaneously, albeit logically, changed to all out support for the "War effort" in the Allied countries. On the trade union front, this was translated into advocacy of a "no-strike pledge" and "production for victory".

C.S. Jackson, president of the Canadian district of the United Electrical Workers (UE) and the most prominent Communist in the Canadian union movement, exemplified his Party's changed position: "Class strife, strikes and lockouts... are the weapons of the Fifth Column of Hitlerism."[18]

The non-Communist leadership in the CCL, though it had supported the War from the beginning, refused to adopt the "no-strike pledge", arguing that it was a policy perhaps acceptable in the United States where federal legislation had been passed legitimizing unions, but totally unacceptable in Canada without comparable legal recognition.

Apart from being identified with the Communist faction in the labour movement, Mine Mill had its own particular history of disputes with the established CCL leadership. In March, 1941, the CCL, then a new organization struggling to give industrial unionism a new thrust in Canada, had offered Mine Mill's Northern Ontario jurisdiction to the United Mine Workers (UMW), because the former union appeared to lack the resources necessary to organize against opponents as tenacious as the mineowners. On behalf of the UMW, John L. Lewis declined the offer.*

Eight months later, Mine Mill's local in Kirkland Lake launched its famous strike against the advice of the CCL leadership. That strike, ending in defeat and the dissolution of the local, also drove the central Congress near to bankruptcy. Following the apparent breaking of Mine Mill's organizing effort in Sudbury in February, 1942, CCL secretary-treasurer Pat Conroy again asked the CIO to reconsider the hard rock mining jurisdiction in Canada. For, with a debt of $35,000, IUMMSW's entire Canadian membership had dwindled to a mere 500.

But Conroy and the other CCL leaders were making a mistake which they were to make repeatedly in their dealings with Mine Mill — they did not consider the always difficult to fathom climate of opinion at the rank-and-file level. Within the year, Sudbury had been organized into the largest local in the entire IUMMSW organization. In recognition, Bob Carlin, its organizer and one of the leaders of the ill-fated Kirkland Lake strike, had been appointed to the international's executive board as the representative of a newly-created Canadian district.

Carlin, as noted, was a member of the CCF. However, in the internal politics of the international union he found himself allied with a faction clustered around its president, Reid Robinson, but dominated by Communist Party supporters. This accommodation seemed to be a strategy of Carlin's to gain autonomy for his Canadian district by rendering unqualified support to Robinson on the international executive board. This tactic was similar to one pursued successfully by George Burt, the Canadian director of the United Auto Workers. Unfortunately, Robinson had few qualms about interfering in the Canadian union, managing to gain acceptance of the "no-strike" policy

---

*CCL secretary-treasurer Pat Conroy came from the Western Canadian district of the UMW.

despite the personal opposition of Carlin (and the CCL leadership) and appointing a well-known Communist Party member, Harvey Murphy, to head a separate western Canadian district.

By 1946, Local 598 was beginning to consolidate its position in Sudbury. An advantageous reclassification had been won in the 1945 contract and the 1946 agreement provided for a ten cent an hour increase in the hourly wage. But at the same time the conflict in the international union was coming to a head. The appointment of Party supporters to organizing, newspaper, and high-level advisory positions throughout the union, a substantial irritant to the non-Communists, had been exacerbated by evidence of corruption at the highest level. With Carlin's support on the executive board, Reid Robinson narrowly escaped investigation for an attempt to "shakedown" a company with a Mine Mill contract.

At the IUMMSW Convention in 1946, an effort was made to ban Communists from union office. Carlin led the Canadian delegation in overwhelming opposition to this move, thus ensuring its narrow defeat. But even within the Sudbury delegation serious divisions appeared. James Kidd, who had succeeded to the presidency of Local 598 in 1945, supported the anti-Communist resolution.

The Convention was followed by a general membership election for the international executive board, with Kidd contesting Carlin's District 8 position. He lost, the Robinson slate was re-elected to the top international posts, and charges were immediately made that the election had been rigged. With eleven locals threatening to secede, Robinson resigned, appointing as his successor Maurice Travis, a prominent Communist who had been expelled from the Steelworkers. An investigation of this affair by the CIO only resulted in a second round of "shuffle diplomacy". Travis resigned, appointing himself secretary-treasurer, Reid Robinson vice-president, and John Clark to the presidency. The August, 1947 Cenvention of the international confirmed this line-up.

In Local 598's June, 1947 election, James Kidd faced Carlin-supporter Nels Thibault in a rerun of the 1946 battle for the presidency. This time Thibault won, at least in part because of an unresolved union trial against Kidd for an alleged breach of union secrecy. In other Northern Ontario Mine Mill locals the outcome of elections was not so favourable to Carlin's position. In an election which was to foretell the divisiveness that would characterize this intra-union feud, Bob Carlin's younger brother Ralph led an anti-Communist slate to victory in the Timmins local.

In the next few years, Ralph Carlin would repeatedly seek aid from Pat Conroy of the CCL and Charlie Millard of the Steelworkers to oust the Communists from the union, and he would eventually organize a secessionist movement within Ontario locals. For his part, Kidd went on to assist the United Mine Workers in two unsuccessful raids, one in 1948 and again the following year, on a small bargaining unit held by Local 598 at the CIL plant in Sudbury. Shortly thereafter, he became a shift boss for Inco.

As if to exacerbate the situation in District 8, Robinson and a number of American Mine Mill organizers who were finding it difficult to operate in the United States because of the anti-Communist Taft-Hartley Act*, decided to move into Northern Ontario. Despite Thibault's and Carlin's dominance in Local 598, a motion to hire Robinson failed at a general membership meeting, leaving the dubious honour of finding a place for the former international president to the Kirkland Lake local.

Obviously these internal struggles within Mine Mill, as complex and heated as they were, were not occurring in isolation from the rest of the world. Throughout World War II, American government strategists had been carefully planning for the "peace".† Theirs was a plan for a massive capitalist expansion — American capitalist expansion — around the world, without the encumbrance of a formal colonial system but rather under the banner of "liberal democracy". Though most of the old European empires were destroyed or severely weakened by the War, one power remained conceivably strong enough to challenge this developing American imperial design — that was the Soviet Union, with its newly acquired "client states" in Eastern Europe and its triumphing allies in China and other parts of Asia. Consequently, the negative complement to the American ideology of "democracy" had to be anti-Soviet and anti-Communist.

Given the about face this represented for a populace well aware of the truly heroic sacrifices of the Soviet people in the Allied War effort, it is not surprising that an intensive "period of adjustment" was necessary to gain popular acquiescence in Canada to this American expansionism. Thus, the "red peril" was given life in American propaganda. But the Communists, too, helped prepare the way for their own eventual isolation with some of their dubious political machinations against other leftists and social democrats.

A case in point is the CP's electoral strategy during the latter part of World War II and its immediate aftermath. The international Communist line called for a kind of "popular front" against resurgent

*This act demanded that elected union leaders sign pledges affirming their anti-Communism.
†See Gabriel Kolko's *The Politics of War*, for a masterful exposition and documentation of this process.

fascism in the Western countries. In Canada, this was translated to a call for an electoral alliance of the Labour Progressive Party (LPP)*, the CCF, and the Liberal Party against the "crypto-fascist" Conservative Party. Despite the role played by the Liberal governments of Mackenzie King and Mitch Hepburn, in 1944 the CP newspaper, the *Canadian Tribune*, was able to describe "Mackenzie King Liberalism" as something close to "what used to be known as communism", while at the same time castigating the CCF.[19]

In a 1945 by-election, the Communist Party endorsed the federal candidacy of Liberal General Andrew McNaughton. In the general election that followed later that year, the Communist-led United Electrical Workers called for the re-election of the Mackenzie King government and George Burt of the UAW ran as a Liberal candidate with Communist Party support.

This support for the Liberals was also part of the bitter contest brewing within the CCL over the role to be taken by the latter's Political Action Committee (PAC). Under the chairmanship of Charlie Millard, PAC had been endorsing the CCF. The Communists retaliated. Even Carlin found himself opposed in the 1945 Ontario provincial election by an LPP candidate, Charles McClure, the editor of Local 598's *Sudbury Beacon*.

Within the CCL, this debate surfaced in the 1946 Convention as a straight power struggle between the Communists and the social democrats. Though they began by attacking the idea of the CCL supporting any political party, the Communist faction suffered from a lapse by one of its militants, who characterized the problem as one of alternatives — either the CCF or the LPP. After a tumultuous eight hour session, the Convention endorsed the CCF.

The intensity of this power struggle as it developed over the next three years is paralleled by its complexity. In 1948, Bob Carlin was denied the CCF nomination for re-election in Sudbury through the intervention of that party's provincial and national leadership. The reason for this denial was clearly his alliance with Communist militants in Mine Mill. He ran anyway, narrowly missing victory as an independent CCF candidate, but in the process ensuring a Conservative triumph. The CCF promptly expelled him.

Outside Sudbury, Mine Mill also seemed to be embroiled in a continuous dispute with the CCL leadership. In May, 1948 the Canadian government deported Reid Robinson, with the support of the Congress executive. By then, Aaron Mosher, the president of the Congress, Conroy, and Millard were intent on purging the Com-

---

*This was the name adopted by the Communists in 1943, following the suppression earlier in the War of the Communist Party for its anti-War activities.

munists, regardless of the implications. Two confrontations later that same year gave them the excuses they were seeking to expel IUMMSW from the CCL. The first involved a raid by the Steelworkers on a Mine Mill local, apparently undertaken at the invitation of anti-Communist dissidents within the union. Refusing to uphold the time-honoured right of any affiliate to demand protection against raiding, the CCL leadership ordered both unions to "back-off" to allow the central to take over the local until the situation could be clarified. Justifiably, Mine Mill rejected this "solution".

The second dispute, though it may seem trivial, provided the final "legitimation" for the CCL's expulsion of Mine Mill. In a union newspaper, Aaron Mosher, in his capacity as president of the Canadian Brotherhood of Railway Employees, was accused of betraying his membership in negotiations with the railways. And at a Vancouver labour gathering Harvey Murphy, the self-proclaimed "reddest rose in the garden", launched into a drunken denunciation of the Congress leadership. Unfortunately for Murphy, William Mahoney, assistant to Charlie Millard (and his successor as Canadian director of the USWA), was present and in town precisely to look for "ammunition" against the Communists.

On August 24, 1948, the CCL executive council used Mahoney's sworn deposition to bar Harvey Murphy from Congress proceedings and at the same time suspended Mine Mill for the newspaper attack on Mosher. Though the union vigorously protested, repudiating and apologizing for the slur against Mosher, its fate was assured by the resolve of the CCL leadership.* Its chances for a reprieve were further minimized by the assistance given by Murphy and the B.C. Mine Mill organization to an attempt by the Communist-led British Columbia section of the International Woodworkers of America to break away from their international.

In 1949, the CCL Convention confirmed the executive council's decision and the suspension became an expulsion.† The Communists were further repudiated with the re-affirmation of Congress support for the two most important instruments of American foreign policy in

*There is a certain irony to this expulsion, based as it was on a breach of the kind of military discipline most often associated with Communist Parties.
†The ideological role of social democracy in the early Cold War period is fascinating. It was the CCF supporters in the CCL who had the political resolve to oust the Communists. In this, they preceded and conceivably influenced the non-Communists in the American CIO. On the electoral level, an argument has been presented that the anti-Communist campaign of social democrat Norman Thomas in a 1948 Congressional campaign in Wisconsin, provided the model for another Wisconsin politician, Joseph McCarthy, for his later career.

the post-War period, the so-called Marshall Plan and the North Atlantic Treaty Organization. That same year, debate on these latter issues marked something of a watershed in the CCF, too, with the expulsion of two MLA's in Manitoba, Berry Richards and Wilbert Doneleyko, for their opposition to the United States' global designs.

MORE THAN JUST A 'BREAD AND BUTTER' ORGANIZATION

It was a mistake to assume that isolation of IUMMSW in national labour circles, even combined with the growing isolation of the Communist Party, would lead to the sudden collapse of its locals, in Sudbury or elsewhere. Once Local 598 had consolidated its position, the days of Sudbury as a corporate fiefdom were over. As early as 1944, Mine Mill had extended its reach beyond the mining companies, securing a contract for the workers in a local CIL plant.

After his election in 1947, Nels Thibault initiated the building of a union hall on Regent Street in Sudbury to serve as a recreation centre. This community orientation probably helped his slate win an overwhelming victory in the December, 1948 election against a list of candidates backed by the local CCF. (However, the turnout, 5000 out of the 12,000 eligible to vote, gave no one cause for complacency.)

In October, 1949, shortly after its expulsion from the CCL, IUMMSW moved to protect itself in Sudbury by chartering General Workers' Union, Local 902. The intention was to organize the service workers of the city, at least in part to prevent the CCL unions from gaining a base from which to attack Mine Mill.

At first bartenders and hotel employees were organized into Local 902, and then taxi drivers. Soon its scope was expanded to include workers in the mass marketing chain stores which had been established in the immediate post-War years. In 1952 a contract was signed with Dominion Stores; and the Loblaws stores in Sudbury were the first in that chain's Ontario operations to be unionized. By 1956, the General Workers' Union held 50 contracts in the Sudbury area. But its successes were not automatic, despite the support of the mine workers. A strike that began against the Metropolitan Stores in October, 1956, ended the following May with the decision of the company to leave town.

Not only on the economic front, but also socially, Mine Mill emerged as a countervailing force to International Nickel. With the success of its first union hall, Local 598 embarked in 1952 on the building of other centres. By 1960, four had been built in the satellite communities of Garson, Chelmsford, Coniston, and Creighton. The latter two included bowling alleys and generally these halls served as "beverage rooms" and centres for athletic and cultural programmes. The Regent Street hall housed a boxing club and all of them regularly scheduled entertainment. The "Saturday Morning Club" featured cartoons and movies for the children; "Family Night" on Sundays usually included a film; and the "Wednesday Afternoon Club" provided babysitting and social activities for housewives. For the more active, there was a women's auxiliary.

Weir Reid, hired as Local 598's director of recreation in 1952, organized the "Haywood Players" (named for "Big Bill" Haywood), providing union men and women the opportunity to study drama, mount productions, and even compete in drama festivals. Teachers were also hired for classes in traditional and modern dance. Special events included performances by Pete Seeger, the Travellers, and other folk, country and western musicians.

The union's summer camp on Richard Lake, opened in 1951 as a day camp, was perhaps its most successful community initiative. For the use of the entire community, it offered picnic facilities, a sports playing field, and craft instruction. After 1953 it provided dormitory facilities which housed over one thousand children over each summer, in one week stints.

Former Mine Mill activist Ray Stevenson explained the motivation for this kind of programming: "Our aim was to provide something that was very badly needed — cultural and recreational development and opportunities in the community . . . we were aiming at deepening the roots of the organization in the community so that we were not just a bread and butter organization. We knew we were under attack from the outside, and we were facing one of the most ruthless companies in the country, so we developed that programme."[20]

Throughout the Fifties some effort was made to maintain this aspect of the union's life, often in the face of considerable difficulties. In 1952, Paul Robeson, American singer and black revolutionary, was prevented from singing at a IUMMSW Convention in Canada by a travel ban imposed on him by the American government.* After this ban was removed in 1956, Robeson gave his first performance outside the U.S. in Sudbury's Regent Street Hall.

In 1954, the union's commitment to the cultural development and well-being of its members and their community was again thwarted when pressure exerted from Washington forced the cancellation of an appearance by the Royal Winnipeg Ballet in the Union Hall.† Prima ballerina Eva von Gencsy explained to the *Toronto Star*: "People told me I had the flu. So I had the flu."[21]

---

*In defiance, Mine Mill organized several concerts featuring Robeson singing from the U.S. side of the border to thousands of Canadians on the other side.
†The RWB was scheduled to appear in Washington after its performance in Sudbury.

THE LONGEST RAID

In January, 1950, following its expulsion from the CCL, Mine Mill's jurisdiction was sold to the USWA for $50,000. This sum was supposed to reimburse the labour central for its expenses incurred trying to organize in this bailiwick itself. In 1948-49, Steel had made a bid to annex IUMMSW Local 637 at Port Colborne. This was the raid which the CCL tried to manage by ordering both contending unions to "back-off", while it chartered the local directly. While that dispute raged, Inco refused to negotiate with Mine Mill and withheld the dues check-off. But in a vote supervised by the newly-created Ontario Labour Relations Board (OLRB), Mine Mill narrowly survived that challenge. In Timmins, it lost two mines to the CCL.

Immediately upon securing the CCL executive's approval, the Steelworkers opened an office in Sudbury. They met so little response that they vacated it in April, 1950, with half a year remaining on their lease. It is important to note that the Toronto *Globe and Mail*, at that time a faithful voice for the mining industry in Canada, the *Toronto Telegram*, and the Sudbury *Star*, all opposed the Steelworker raid on Mine Mill, casting very considerable doubt on the importance of the "Red Peril" in this struggle.

By 1951, Mine Mill seemed to have been stabilized. However, the battle had taken its personal toll and Bob Carlin was no longer administering District 8 effectively. In August, he was replaced by Nels Thibault and Mike Solski became 598's president.

A group in opposition to Thibault and Solski remained cohesive throughout 1952, prompting the leadership to dismiss a number of stewards for their activities. Despite this high-handedness, Solski easily beat Bain McKelvie, 4765 to 3332, in the November election that year. Following this failure, the opposition disintegrated and the incumbents were acclaimed in 1953.

In 1954, James Kidd reappeared on the scene. Demoted from his position as shift boss, he returned to union activity with a vengeance. However, his opposition to the Mine Mill leadership was easily construed as nothing more than a raid, and he was discredited. A union tribunal expelled him for life, ten of his supporters were suspended for a year and fined $50, while a number of others were fined $25. The penalty for Kidd was inconsequential; by then he had become a staff representative for the USWA, who re-opened an office in Sudbury.

In the 1955 and 1956 local elections, the Solski slate was returned unopposed. Opposition within the union evoked a "machine" response from the leadership, which automatically equated dissent with attack from without. A good portion of the rank-and-file assumed a passive attitude to the local.

On the political level, the Sudbury Mine Mill met little success. An attempt to elect a union slate to City Council in 1954 was a dismal failure. Federally, the Local's Political Action Committee assumed a "non-partisan" guise which effectively resulted in support for the Liberal Party. In 1955, Thibault and Carlin contested the provincial seats of Sudbury and Nickel Belt as "farm-labour" candidates. The *Mine Mill News* felt compelled to exhume Samuel Gompers' slogan, "reward our friends and punish our enemies", in an unsuccessful effort to drum up enthusiasm.

On the whole, the mid-Fifties were relatively quiet for Mine Mill in Sudbury. So much so that a referendum on a dues increase in 1958 attracted only 1566 voters out of 13,000 eligible. And the development of the uranium mines around Elliot Lake had shifted the battleground with the Steelworkers.*

THE STRIKE OF 1958 AND THE DEFEAT OF MINE MILL
The failure of Mine Mill to collapse or to be deserted by its membership in 1949 and 1950 indicated to some the lack of a strong "ideological" alternative to its kind of unionism. "Adult education" was one way to develop and motivate the leadership necessary to depose the established leaders of Local 598.

In the mid-Fifties, the University of Sudbury, a Catholic college within the newly-created Laurentian University, formed the Northern Workers' Adult Education Association. Alexander Boudreau, an economist with experience in the Nova Scotia co-operative movement, was hired to educate workers to a "new conception of unionism" — one capable of battling the "Communism of Mine Mill", in fact, one that promoted a larger struggle against Communism as a major task of unionism. Among the materials used was literature from the "Christian Anti-Communist Crusade", a right-wing, fundamentalist American group. Even James Kidd found this emphasis disturbing.

Boudreau's "Co-operative" ideology, pioneered in Catholic unions around the world,† envisioned harmony between workers and management in pursuit of such "common goals" as efficient production. In predominantly Catholic Sudbury, such an approach was assured of an audience.

It was not surprising that Laurentian's Board of Governors wel-

---
*This was to become a financially disastrous campaign in which Mine Mill was only able to win one unit, which it lost back in 1959.
†Though Catholic unions were originally formed to counter the threat of "subversive unionism", many of these confessional unions have become secularized over the past three decades and some are now a challenge to the order they once defended, eg. the CSN in Quebec.

comed the introduction of corporatism* to Sudbury. Its chairman was Ralph Parker, a vice-president and operations manager of Inco, while the President of the University, Emile Bouvier, was a powerful Catholic who had been involved in "red-baiting" during the famous 1949 asbestos strike in Quebec.

But this emphasis on "adult education" could only provide a potential leadership which would still have to ground its appeal in the more mundane reality of union politics. The opportunity to do that came in 1958.

Beginning in 1957, depressed markets combined with the end of the U.S. government's stockpiling to put International Nickel in the position of having huge inventories. As well as the investment tied up in inventories, at a time when there was a glut of nickel, the Company was developing a nickel deposit at Thompson in Northern Manitoba.

In March, 1958, Inco laid off 1000 workers in Sudbury and 300 in Port Colborne. This was followed in April by a further cutback involving 300 employees. Mine Mill proposed 30 hours work for 40 hours pay to alleviate the effects of the layoffs. Inco responded in June by cutting back to a 32 hour work week at the established wage, effectively reducing incomes to the level which had prevailed in 1951.[22] Of course, this "belt-tightening" happened to coincide with contract talks between the Company and the local union.

From the Company's perspective, a strike at Sudbury would be a convenient way to cut both inventories and costs. Therefore, its first offer was for a two-year contract with no wage increase. Even Ontario's Tory Attorney-General Kelso Roberts felt compelled to chastise the Company for its intransigence. But Inco refused to budge, claiming hard times.†

On September 13, 1958, the members of Locals 637 and 598 appeared to spring the management's trap, in Sudbury 10,662 out of 12,887 voting to strike.‡ Inco's Canadian operations were completely shut down for three months but it still refused to make any major

---

*Corporatism theorizes a society composed of distinct and recognizable corporations, similar to classes; however, their interests can be reconciled through the corporate state.

†Conveniently, John Diefenbaker's new Conservative government in Ottawa was trying to encourage a policy of "wage restraint".

‡This case poses one of the classic dilemmas of unionism. The preparation for bargaining seems exemplary. Research Associates of Montreal was commissioned to investigate the economics of the situation, providing an analysis that was later proved correct. The Local scheduled seminars for stewards to discuss the negotiations well before they got under way. Yet it could not possibly win against a Company that had nothing to lose.

concessions. On December 22, Mine Mill ratified a three-year contract providing for annual increases in the hourly rate of one, two, and three percent.

The failure of the strike was a key factor in the turnover in the leadership of Local 598 followed in March of 1959. Mike Solski and his supporters were ousted by a slate of "reformers" headed by Don Gillis, the reeve of Neelon-Garson township. Solski, in fact, received 3830 votes, 98 more than he had needed to win in 1957; but Gillis' "Committee for Democratic Leadership and Positive Action" was able to mobilize many of those who had been passive, registering 5629 supporters for their leader.

Gillis and many of the others in the new regime on Regent Street were graduates of Boudreau's leadership training courses. As the Adult Education Association had used materials supplied by the USWA and the Canadian Labour Congress (CLC)*, it was not surprising that this new leadership should pledge to bring the Local back into the "mainstream of organized labour" (the CLC). Quickly the lines were drawn between the National Office of Mine Mill and the Local.

One of the first actions of the new executive was to fire without notice the Local's recreation director, Weir Reid, a Communist who had been responsible, in large measure, for Mine Mill's community programming. This sparked an uproar, as Reid had long been a controversial figure — detested by the City's elite but much admired by those who had benefitted from Local 598's community orientation.

Undaunted, the "reform" team then commissioned Allistair Stewart, a CCF activist from Manitoba on the staff of the United Packinghouse Workers of America, to audit the Local's financial records. Although he was unable to unearth any fraud or theft, Stewart discredited the Thibault-Solski leadership by creating the impression that it had been irresponsible if not dishonest. He emphasized a loan made by the Local to the National Office for the construction of a hall in Elliot Lake, at a time when 598 was in a tight financial position — during the 1958 strike. And in an unauthorized and provocative move, Stewart released his "findings" to the press. What the Toronto *Telegram* was to dub "the bitterest intra-union fight Canada has ever seen"[23] had begun in earnest.

Reid's partisans, in the meantime, organized an illegal Local meeting at which they reinstated him. The executive laid charges against the three "Old Guard" trustees who had convened that meeting and held their own a few days later. At it the recreation director's firing was

*In 1956, following the merger of the AFL and the CIO in the U.S., the TLC and the CCL joined to form the Canadian Labour Congress.

confirmed. In a scuffle that followed, Reid and one of his supporters allegedly "knocked-out" Ray Poirier, one of Gillis' closest collaborators. Another trial was immediately scheduled.

The issue that remained the focus for the whole debate, however, was the union's relationship to the CLC. Though both the National Office and the Local's new executive appeared to agree that a general membership referendum should be held to resolve the matter, the CLC discouraged the idea. The latter's leadership appeared to have its own agenda for any reconciliation and it appointed a staff representative, James Robertson, to advise Gillis on the path to follow.

Communism, or at least Mine Mill's tolerance of Communist Party members, soon emerged as the major apparent block to IUMMSW's return to the "House of Labour". Gillis began to use the union newspaper to "develop" this issue. And at the National Convention of Mine Mill in the fall of 1959, Local 598 submitted a resolution which would have barred from union office anyone " . . . associated whatsoever with the Communist Party or any group which expounds or promotes any doctrine or any philosophy contrary to or subversive of the fundamental principles and institutions of the democratic form of government of Canada." The motion was defeated.

To contest the November, 1959 Local election*, Nels Thibault resigned as National Union director. The ensuing campaign set the tone for the near-continuous debate of the next two years. Alexander Boudreau and his allies in the churches waded into the fray, characterizing it as a "last ditch fight between Christianity and Communism". A Catholic Social Life Conference was held in Sudbury, according to Mayor Joseph Fabbro, because the City had become a haven for Communism, in need of expiation. A headline in the *Star* told its readers that they "Must Prove Sudbury Not Communist Area". One Thibault supporter recalled: "I used to go to church just to hear what they were going to say against me."[24]

Boudreau, a master of the media, had orchestrated to perfection his general crusade against Communism. He used every platform available, lecturing in churches, schools, and service clubs throughout the area, pontificating in the press and in his regular television appearances.

> Boudreau "used to have a drawing board in the TV there . . . you weren't only hearing it, you were seeing it . . . that boob tube is the worst thing for brain-washing people I ever saw."[25]

*The election in which Gillis had first been elected, in March, 1959, had been postponed from the previous year because of negotiations and the strike; thus, he served only a half term before facing re-election.

Thibault was accused by Boudreau of using an alias and by Ray Poirier of speaking Polish. The implication was that he was an agent trained in Eastern Europe. A series of seven articles appeared in the Toronto *Telegram* based on interviews with Poirier, it was later revealed. These painted a lurid picture of prostitution and blackmail and the teaching of children to inform on their parents. A typical headline read: "Ontario Reds Recruit Seven Year Olds".*

With this kind of continuous "unofficial" campaigning, Gillis could afford to run on a moderate platform. On paper, there was little to choose between his and Thibault's programme. But Gillis was easily re-elected, 7221 to 5903, in an 81 percent turnout. Neither side was yet ready to back down, though.

In 1960, the CLC rejected the Canadian IUMMSW's application to join. But it continued its assistance to 598, even conducting the Local's stewards' schools. The factional fighting continued, each side controlling a newspaper. Membership meetings became increasingly unmanageable and more trials were held. Those sympathetic to the National Office were "red-baited", while those siding with the Local executive were accused of conspiring to assist a Steelworkers raid. The National Office attempted to devise a national re-structuring that would isolate Gillis and reduce the influence of 598 across the country. The Local found some pretext to withhold its per capita tax from the National.

The campaign for the June, 1961 Local ballot was a continuation of the preceding one, though the planting of two bombs in the cars of Gillis supporters may have been an indication of the frustration and desperation of the Thibault forces. Again Gillis and his slate were returned, with a slightly reduced margin of 1000 votes. The following month, the National Office suspended 598 for its dues default; and the CLC recommended to the Local that it join the Congress by affiliating to the USWA. Gillis' response was unequivocal — he invited the Steelworkers to help in preparations for up-coming contract negotiations with Inco.

Near the end of August, the National Office imposed a trusteeship on the Local and appointed William Kennedy its administrator. With Gillis out of town, and armed with a court injunction, Kennedy took possession of the Regent Street ofice. A counter-siege ensued, with the protagonists using broken beer bottles and fire hoses. Eventually, the Riot Act had to be read and the crowd dispersed by the police. A few days later, on September 5, the Ontario Supreme Court overturned the previous injunction and the Gillis forces regained possession of all 598's facilities.

*The *Telegram* later settled out of court a libel suit initiated by Weir Reid.

By this time, Steel's raid on Mine Mill had become overt. A rally in the Sudbury area for September 10 was scheduled to be addressed by Claude Jodoin, president of the CLC, and by William Mahoney and Larry Sefton of the USWA. But by 9:30 that evening the building had to be evacuated, after tear gas was used to break-up fights in the crowd of 8000. Ten people were hospitalized. On September 15, Mahoney regained the initiative — he announced that Bob Carlin had joined the Steelworkers.

In November, the Local executive suspended monthly membership meetings because they had become totally disorderly. Mine Mill National, in the meantime, was finding its attempts to discipline Gillis thwarted by civil court injunctions. Thus, a group openly committed to affiliation with Steel was free to use the facilities of Mine Mill's largest local in the service of their cause. The Mine Mill loyalists were left in a very weak position.

Inco was quick to take full advantage of the situation. When 598 discontinued membership meetings, the Company refused to recognize stewards or process grievances. Similarly, it refused to bargain for a new collective agreement. Eventually, Sudbury's mine workers would go 19 months without a wage hike.

In December, the Ontario Labour Relations Board accepted a USWA application to be certified the bargaining agent for Sudbury Inco and Falconbridge workers. The deciding vote was set for February 27 to March 2, 1962.

Right from the start, this final campaign was not confined to Sudbury. The first outside influence was the certification vote at the Port Colborne Inco refinery in early December — Steel defeated Mine Mill, 1033 to 763. The other "external influences" were of more dubious relevance. George McClelland, Deputy Commissioner of the RCMP, gave a speech in support of Gillis at a conference in Halifax, as did federal Justice Minister Davie Fulton in the House of Commons. In the United States, "coincidentally", Secretary of Labour Arthur Goldberg and Attorney-General Robert Kennedy denounced the activities of the Communists in Mine Mill and the Subversive Activities Control Board accepted the testimony of six Steelworker officials on this subject. Just before the final vote, the Conservative government of the province appointed Don Gillis to the Ontario Economic Council.*

Mine Mill's only recourse was to the immediate community. In January, it organized a women's committee and tried to involve entire

---

*Gillis was also sent by the federal government as a labour representative to a NATO conference in 1962 and later that year he ran as a Conservative in the federal election, pledging to have the Communist Party outlawed.

families in its defence. It put forward the slogan "a union without the women is only half a union".

When the ballots were tallied in March, it became clear that almost every worker at Inco had voted. The Local was nearly evenly divided — 7182 for Steel, 6915 for Mine Mill. As the 168 abstentions counted for Mine Mill, according to OLRB rules, the USWA was victorious by a mere 15 votes. A request for a recount was denied by the Board and on October 15, 1962, the Steelworkers' Local 6500 was certified. However, the vote at Falconbridge, which had taken place simultaneously, was invalidated because of 50 apparent forgeries favouring the USWA. Steel discreetly withdrew from that contest. But in Thompson, Manitoba it trounced IUMMSW*, 1226 to 352. Finally, the USWA had established itself as the main union in the Canadian mining industry. Inco welcomed its new adversary with a lay off of 2200 workers late in 1962.

They had right-wing union fighting left-wing union and the boss came right down the middle and cleaned everyone of them out.[26]

Mine Mill attempted a comeback raid in 1965 but lost by 2000 votes. In 1967 the two antagonists reached an accommodation and the smaller union was absorbed, except for a fragment of Local 598 which represented the workers at Falconbridge. It opted for independence and remains to this day a remnant of what had been the largest miners' union in Canada.

The final irony is the fate of the main Mine Mill leaders. Even "the reddest rose", Harvey Murphy, was readily incorporated into the Steelworker apparatus. And Nels Thibault climaxed his later career in the USWA with election in 1976 as the "official candidate" to the presidency of the Manitoba Federation of Labour.†

THE AFTERMATH

The memory of the bitter three month strike of 1958 did not fade during the period of division and uncertainty in Sudbury. In July, 1966, during negotiations in which Inco was again proving intransigent, a wildcat strike broke out at the Levack Mine. The workers at the Iron Ore Recovery Plant followed and the next day all 17,000 Sudbury

---

*Mine Mill was an easy target in this new mining town because it had not established a functioning local office, but was conducting all important business from afar.
†As the Steelworkers are the largest union in Manitoba, they tend to provide the leadership for the MFL.

workers, plus those at Inco's refinery in Port Colborne, joined the "illegal" walkout.

In Sudbury the plant gates and mine heads were effectively blocked by picket lines, vehicles, and even large boulders. A truck carrying food supplies was overturned and rolled down a hill. The Company rented a transport helicopter to move "scabs" and supplies into its premises. Telephone and power lines were cut. A few days after the strike had begun, a large squad of Ontario Provincial Police was moved into town, amid rumours that the strikers were armed with clubs.

This tense situation was defused somewhat when the USWA agreed to let supervisory personnel into work. After three weeks the wildcat ended and the negotiations were resumed. But by mid-September, Inco's workers at both Sudbury and Port Colborne were again "hitting the bricks", this time "legally".

The Company was in no position for a long strike. Nickel demand was strong due to the needs of the American war machine in Vietnam. After a mere three days, Inco doubled the wage increase it had offered the previous month, in essence agreeing to union demands. The members of Local 6500 voted 57 percent in favour of settlement. In some sense, this skirmish was a preliminary round for the negotiations of 1969.

Nickel markets continued to be strong throughout the life of the 1966 contract. The Vietnam War continued to escalate, and the economies of the Western world were generally expanding. Though Inco had finally decided to diversify and increase its sources of supply around the world, none of its new mines was even under construction, let alone near production. Members of Local 6500 were justified in thinking that their position was strong. But the Company, always with an eye to maintaining its authority, refused to accept the union's demands. The result was a four month strike.

The final settlement, when it came, was an unqualified victory for the workers. Wall Street was as infuriated by this "blackmail" as were the families of Sudbury jubilant about their victory and final revenge for 1958.

WHO DUG THE MINES? 61

Inco in Guatemala

The Cold War concealed the unprecedented expansion of American-based transnationals which followed World War II. This direct use of state power to further corporate interests was symbolized by a meeting of the key members of President-elect Dwight Eisenhower's cabinet in 1952. Appropriately occurring on board a U.S. warship, it was intended to plan the new administration's foreign policy, or in the language of the day, to decide "how best to combat Soviet-dominated Communism throughout the world".[1]

In attendance were John Foster Dulles of Sullivan and Cromwell, longtime counsel and executive committee member of International Nickel; George Humphrey, for twenty-three years President of the Hanna Mining Company; and Charles Erwin Wilson, President of General Motors. Respectively, they were appointed Secretary of State, of the Treasury and of Defence.

Though this coincidence of corporate and political power generally facilitated the expansion of the transnationals around the world, its impact on Inco was not so straightforward. For, as noted, one of the early conclusions reached by the planners of this expansion of American industrial might was that the supply of nickel was not growing fast enough to match the projected demand. Therefore, the American government had decided to encourage production by subsidizing some of the smaller nickel producers, including Falconbridge, Freeport Sulphur, and Sherritt-Gordon Mines. Thereby, Inco had been prodded to expand its productive capacity in order to defend its market position.

Some of the Company's growth occurred within Canada, notably in Northern Manitoba, at a site called Thompson after Inco chairman Dr. John F. Thompson. By 1960, this $400 million complex was approaching full production. But the economics of mining various types of ore was also drawing attention away from the deep sulphide deposits of the Northern Hemisphere to the lateritic nickel of the tropics.

In 1956 the Hanna Mining Company of Cleveland had secured a concession on the shores of Lake Izabal in Guatemala, from the military government of Colonel Carlos Castillo Armas. Unable or unwilling to develop these deposits by itself, Hanna turned over control to Inco in 1960 with the incorporation of a new company, Exploraciones y Explotaciones Mineras (Exmibal). The Cleveland company retained only 20 percent of the equity in Exmibal, while Inco took the remaining 80 percent.

Hanna had secured these mining rights only two years after a CIA-inspired coup in Guatemala and was probably seeking to reduce its own risk by involving the larger company in the development. The cause for this caution, then, is to be found in Guatemala's recent history.

GUATEMALA'S LIBERAL REVOLUTION: 1944-1954

In 1944, the corrupt Ubico dictatorship was overthrown by a coalition of nationalist army officers and intellectuals with massive popular support. A year later, Juan Jose Arevalo was freely elected President of Guatemala.

Arevalo styled himself the Latin American Franklin Delano Roosevelt. His government adopted a constitution guaranteeing free speech and freedom of the press. Trade unions were legalized and the right to strike was recognized. Forced labour was abolished and peasant organizations were allowed to operate openly. The U.S. State Department, forever vigilant in such matters, estimated that within a short time, the National Peasants' Federation had organized 215,000 people.[2]

The 1950 election victory of Jacobo Arbenz confirmed this modernizing trend. With the support of the organized workers and peasants and some elements of the middle class, Arbenz was determined to lead his country in a break with its 400 year history of dependence on external centres of power. His goal was to be accomplished through the construction of a modern, independent capitalist economy.

The first step in this strategy had to be a sufficient redistribution of wealth to create a viable internal market for Guatemalan goods and services. Given the overwhelming agricultural nature of the economy, this entailed a substantial land reform. Only in this way could the peasantry attain the purchasing power necessary to participate meaningfully in a market economy.

Decree 900, enacted in 1952, legislated the redistribution of those large estates which were not being used productively. The landowners were to be compensated.

It wasn't we who were in the government; it was Colonel Arbenz, who was merely a friend of ours.
— a Guatemalan peasant[3]

Among the lands to be redistributed were those of the American-based United Fruit Company (UFCo). Though it owned 550,000 acres in Guatemala, more than held by half the country's population, UFCo accounted for only 15 percent of agricultural production.[4] The government expropriated 387,000 acres of its holdings, calculating the compensation according to the value which had been declared for this land in tax returns. This enraged UFCo's owners, who claimed that they had deliberately undervalued the company's assets in conformity with "generally accepted accounting principles".

In the first months of 1954, over 100,000 landless families benefitted from the agrarian reform.[5] In some instances the peasants independently took the initial steps toward the breaking up of the *fincas* (estates) on which they had laboured for so many years.

At this point the entire apparatus for the direct defence of American corporate and geopolitical interests in the area was activated.

KEEPING GUATEMALA SAFE FOR INVESTMENT

In June 1954, the United States smashed Guatemala's ten year experiment with reform. The events surrounding this intervention have been thoroughly documented.[6] The State Department, headed by John Foster Dulles, and the CIA, directed by his brother Allen, conceived, organized, and financed a coup. The "liberation forces", as the U.S. State Department called them, consisted of Guatemalan exiles and foreign mercenaries operating from nearby Honduras and armed with American arms. A decisive factor in the invasion was the cover of P-47 aircraft piloted by Americans.

In the 1930's, John Foster Dulles had helped the United Fruit Company negotiate contracts with the Guatemalan government. His brother Allen Dulles had later become President of UFCo.

In 1954, they were both in good positions to defend the interests of their corporate mentor.[7]

On July 3, 1954, Carlos Castillo Armas, a graduate of the U.S. Army Command and General Staff School at Fort Leavenworth, Kansas, landed in Guatemala City aboard the personal plane of American

Ambassador John Puerifoy.* Five days later he was elected president by a five man junta. The popular support enjoyed by Arbenz did him little good; an unarmed people could not defend the reforms he had initiated. And the Guatemalan Army remained "neutral".

The Castillo Armas government returned the expropriated land to the UFCo and the other landlords. Peasant organizations and labour unions were outlawed and the opposition press was suppressed. Cooperatives were eliminated and teachers who had gone to work in the rural areas were dismissed, since their basic adult literacy programmes were considered a "Communist threat".

The U.S. justified its intervention in Guatemala by portraying Arbenz as a "fellow-traveller" or a Communist. And Communism was characterized as an irrational force, therefore difficult to combat by normal means.

Ambassador Puerifoy: *Communism, in my opinion, is a religion, Mr. Feighan. I don't think there is any doubt about that. Anyone who thinks it is a theory...*
Congressman Feighan (Ohio): *It is a religion. It was originated in hell, with the assistance of Satan and all the evil forces.*
Ambassador Puerifoy: *That is a better definition than mine.*
— testimony to a U.S. House of Representatives Committee[8]

Castillo Armas was only re-establishing an older pattern of government in Guatemala, one which continues to this day. Though it rests on the confidence of the transnational investors, it perpetuates the poverty and underdevelopment firmly institutionalized in the country's early colonial period.

DEPENDENT DEVELOPMENT:
THE COLONIAL BACKGROUND

When the Spanish *conquistadores* came to Guatemala in 1524, they were seeking gold. Instead, they encountered the agricultural civilization of the Mayas, a people who had not developed the wheel or metal tools but had attained a high level of sophistication in their architecture and craftwork, and in their mathematics and astronomy. The Spanish attributed these achievements to the Devil and failing to find gold, embarked on the destruction of the people and their culture.

*Puerifoy seems to have been something of a specialist in organizing this sort of operation. He was posted in Greece during the civil war there in 1949 and after his Guatemalan sojourn, he was posted in Thailand, a country convenient to the "troubles" in Southeast Asia — Vietnam, Laos, Cambodia, and Indonesia.

> *Battle, conquest, slavery, and tribute, were hardships that (the Mayans) had known, but never to the genocidal proportions effected by the white man.*[9]

Two thirds of the native population were slaughtered, as the Spaniards transformed these lands into a settler-plantation economy geared to the export of wealth to Spain.*

As the Spanish Empire declined in the nineteenth century, political independence became attractive to the merchants and landlords of Guatemala. An independence movement arose, gathering strength from the rebellions of the indigenous peoples. In 1821, an alliance of merchants and plantation owners, backed by the might of the still ascendant British and French Empires, declared independence from Spain.

Trade increasingly shifted away from Spain. First, timber to Britain became primary, but then coffee, financed by German capital, emerged as the staple export. By World War I, 90 percent of the country's foreign exchange derived from this one cash crop.

In 1903, the Boston-based United Fruit Company gained the first substantial American foothold in Guatemala — a lucrative land concession on which to grow bananas. One of those who helped win this agreement was William Van Horne, famous in Canadian history for his role in building the Canadian Pacific Railway. "We asked for everything we could think of, and we got all we asked for," Van Horne bragged.[10] In exchange for completing a 60 mile stretch of railway, UFCo received 200 miles of existing track, as well as 170,000 acres of prime agricultural land.

From this beginning, American capital grew to control the country's economic and political affairs. In the process the homeland of the Mayas became the prototype of what North Americans condescendingly term a "banana republic". The reality of that status for the peasants of Guatemala is an annual migration from their rocky subsistence plots in the highlands to the plantations of the fertile coastal plains. There they work to produce the main export crops — coffee, cotton, and sugar.*

---

*The *conquistadores* rationalized the theft of the Mayan land with ease. Since all land came from their God, and his earthly envoy, the Pope, had granted it to the king of Spain, who in his turn had passed it on to them, this land was theirs by divine sanction.

*Given the widespread notion that countries in Latin America like Guatemala are "feudal", hence their backwardness, it is worth emphasizing the real nature of their economies. They are fully integrated into the world economy, depen-

INCO'S GUATEMALA ADVENTURE

When Hanna Mining invited Inco's participation in the Lake Izabal development, the vehicle for co-operation became a Canadian-based holding company called Explorer Metal. Through this firm, the operating company, Exmibal, is owned by the two principals.

By the summer of 1962, the new company was technically ready to begin actual mine development. However, the parent companies felt the need for considerable "political work" before production could be assured at the desired rate of profit. What followed, according to a study of Exmibal carried out by the Faculty of Economics at the University of San Carlos in Guatemala City, were ". . . devious operations and political pressures . . . utilized in order to take possession of this non-renewable wealth of our country."[11]

That there was no satisfactory legal code governing mining was both an impediment and an opportunity for Exmibal. The existing law dated from the Ubico dictatorship of the 1930's and lacked provisions appropriate for the regulation of the kind of modern mining and metallurgical complex being proposed for Lake Izabal.

The company began by pressuring the Ydigoras government to devise a new mining code. It appeared to be making progress when Ydigoras was deposed in a coup d'état staged amidst a storm of popular protest over corruption, electoral fraud, economic chaos, and increased but ineffective repression. Colonel Enrique Peralta Azurdia, leader of the new junta, immediately suspended the constitution. Sensing opportunity, Exmibal's management moved swiftly to encourage promulgation of comprehensive mining legislation and itself hired a Peruvian mining engineer, Emilio Godoy, to draft such a code. In April, 1965, the Guatemalan Congress passed this draft as Decree 342 and four months later Exmibal received the rights to the Niquegua nickel deposits in the hills overlooking Lake Izabal.

Significantly, the mining code was enacted between the suspension of the country's constitution in 1963 and the institution of its successor in September, 1965. Hence, Decree 342 seemed to be of dubious legality. However, the new constitution did not include the previous one's requirement that all mineral exploitation agreements be ratified by Congress. It also contained several articles directly from Godoy's draft mining code. So Exmibal seemed to have ended up influencing even the new Guatemalan constitution. Still it was not satisfied.

As do many other countries, Guatemala offers tax incentives to

---

dent on the production of export crops, with the result that they often have to import food. This is certainly capitalist underdevelopment; it is not feudalism.

companies investing in manufacturing and secondary industry. The point of this sort of strategy is to create permanent jobs by diversifying the economy away from the mere export of raw materials. Classification as a "transformation industry" can mean reduced tax rates for periods up to ten years. Exmibal wanted to have its strip mining operation categorized as a transformation industry.

The first request for this preferential status, made in 1967, was rejected outright by the Minister of the Economy Roberto Barillas Izaguirre. He noted that Exmibal's project came under the mining code and that only non-metallic mineral operations could qualify for the transformation status. Not content with this refusal, the company sought legitimation for its claim by commissioning the Central American Research Institute for Industry (ICAITI)* to do a special study. ICAITI, as expected, recommended to the Guatemalans that they grant Exmibal its request. In May, 1968, with a new minister of the economy in office, the government declared the company eligible for special tax treatment.

Yet that was still not all that Inco had been seeking. The parent company was concerned that its offspring be able to repatriate profits at will to head office. Ever since the invasion of 1954, the pro-American regimes of Guatemala have routinely granted foreign corporations approval for profit remittance. Nevertheless Inco directed Exmibal to avoid the legal stipulation that export earnings be deposited in the host country's central bank before being sent out of the country. Conceivably, such a provision might be used by an unfriendly government in the future.

The rationale on which Exmibal based its request for exemption was its need to accumulate a capital fund outside the country with which to amortize the foreign loans for the investment. On March 29, 1968, the government's Monetary Board ruled, in Resolution 5727, that companies with large foreign debts could deposit funds outside the country. Inco had won all of the major concessions that it had been seeking from the government.

Reaching agreement in the Central American Republic is not easy. Laws governing foreign investment are lenient but when it comes to a contract, Guatemalan

*This is the industrial research arm of the Central American Common Market, an American-inspired and financed body, which seeks to integrate the economies of the Central American nations in order to facilitate large scale foreign investment projects. In a tariff-free common market, it is possible for transnationals to centralize production and thereby decrease costs while maintaining access to the largest possible market.

> *officials tend to nit pick. Thus, there are always lengthy delays.*
> — *Toronto Star*, January 15, 1971

The nickel giant required one more substantial test of the government — it wanted an indication of firm control over a political opposition that might jeopardize foreign investment.

The hills around Lake Izabal were the base for left-wing guerrillas throughout the mid-Sixties. Both the Fuerzas Armadas Rebeldes (FAR) and the Movimiento Revolucionario 13 de Noviembre (MR 13) had established a strong base of support among the peasants of the provinces of Zacapa and Izabal. But just as the United States had made Guatemala "safe for democracy" in 1954, so Colonel Carlos Arana Osorio made even these provinces safe for Inco in 1968. In that year, Arana's American trained and armed soldiers launched a "pacification campaign" to destroy the guerrillas. By the end of the decade over 3000 Guatemalans had been killed in this exercise, many of them peasant supporters of the opposition.

> *International Nickel went to the underdeveloped and politically explosive area of Guatemala — headquarters for Guatemala's leftist guerrillas until separate efforts of the Armed forces and right-wing terrorists pretty well eliminated them in the course of 1968 — because that's where the nickel is.*
> — *Business International*, 1969[12]

On the strength of his military prowess, Arana ran for the presidency in 1970. "If I am elected, all Guatemala will be like Zacapa," he pledged. After his victory, Exmibal was finally ready to proceed and in February, 1971 an agreement was signed. According to Inco, it brought "together certain conditions contained in the laws of Guatemala and other conditions mutually agreed upon ..."[13] A complex capable of producing 60 million pounds of nickel annually was to be built at a cost of $250 million.*

The company agreed to pay half the usual 53 percent tax on mining operations for the first five years of production and three quarters the usual rate for the following five years. As it is likely that the richest ore will be mined during the first decade, taxes will be lowest at the time when profits should be highest.

Perhaps the most highly touted aspect of the agreement is the provision for the participation of the government in the project

---

*The completed project cost $224 million, but is only capable of producing 28 million pounds of nickel.

through the acquisition of up to 30 percent of Exmibal. This equity will accrue in lieu of taxes.

This state involvement in the enterprise seems to have a dual purpose. First, it may be some effort to appease those in the country who see the project as a giant sell-out of non-renewable resources, with little or no benefit to the average Guatemalan. Second, it may have the effect of "vaccinating" the company against full nationalization. But although the government may eventually own 30 percent of Exmibal's stock, it will have little control over its management. Most of the board of directors will continue to be appointed by those interests holding the controlling block of shares — Inco. While the stock may pay dividends, the amount will be voted by the board.

> *The military will continue to rule in Guatemala for the foreseeable future . . . It is the only base of stability, really. It will rule even with a civilian government in power . . . the political prospects are good . . . one of the best prospects in terms of realism and pragmatism regarding foreign investment.*
> — an Inco executive, 1973[14]

Exmibal's status within Inco's internationally integrated operations may be expected to make it difficult to calculate the real profitability of the Guatemalan complex. With control of all stages of production, marketing, and sales, Inco can take its highest profits at the stage and in the country where the taxes are lowest. Further, the fees for the management, technical, and marketing skills which Inco "sells" to its subsidiary can be used to raise the costs and reduce the profits on which Exmibal's dividends and taxes will be based. The same sort of manipulation of costs for transportation and further processing of Guatemalan ore can easily be envisioned.*

As Inco controls virtually all aspects of the project, the Guatemalan state seems to be in a poor position to gain many benefits for its citizens from the exploitation of their land's resources, regardless of formal agreement of "participation".

FINANCING EXMIBAL: CORPORATE WELFARE ABROAD
The skill displayed in manoeuvering within Guatemala was matched by Inco in securing the financing for the project from various international sources.

In 1971, the year in which Exmibal signed its agreement, the Popular Unity government of Chile nationalized that country's copper

---
*Transnationals naturally dispute this argument. But it would seem that only the strongest nation-states have the power to combat this practice with the kind of anti-trust action that the U.S. threatened to use against Inco in 1946.

industry, including within the expropriation the American-owned Anaconda and Kennecott companies. Suddenly private investors were wary of mining projects in Latin America. According to the *Financial Times* of London: "International Nickel's Exmibal subsidiary (was) having difficulty obtaining financing for its $250 million nickel mining project in Guatemala following Chile's expropriation of the U.S. copper companies."[15]

To complete matters, Inco had an untypically poor profit performance in 1971. But the still fresh memory of a bitter four month strike in 1969 at Sudbury acted as an incentive for the Company's management to proceed with the Lake Izabal project as quickly as possible. For that strike had cost millions of dollars in lost profits and ultimately ended with a union (USWA) victory.* Inco turned to the International Finance Corporation (IFC), a unit of the World Bank, for assistance.

The IFC exists to help corporations with their investments in the Third World. It is controlled by the industrialized, capitalist nations, over half the voting power being held by the United States and the members of the European Economic Community.

According to its first Article of Agreement, the IFC should: "Further economic development by encouraging the growth of productive private enterprise ... In carrying out this purpose, the Corporation shall:

    i. in association with private investors, assist in financing the establishment, improvement, and expansion of productive private enterprises ... *without guarantee of repayment by the member government concerned* in cases where sufficient private capital is not available on reasonable terms.

    ii. seek to bring together investment opportunities, domestic and foreign capital, and experienced management

    iii. seek to stimulate and help to create conditions conducive to the flow of private capital, domestic and foreign, into productive investment in member countries." (Emphasis added.)

Because Inco had not received an investment guarantee from the Guatemalan government, the IFC was an ideal institution to assist in the capitalization of Exmibal. As well as direct financial support, IFC's involvement provided much needed legitimacy for the project in the cautious international investment community. A condition of this endorsation, however, was a reduction in the size of the proposed complex. But even the cost of this new smaller-scale project was pushed back up over $200 million by the relentless inflation of the Seventies.

---

*Fortune Magazine lamented this settlement in the following terms: "In effect the union held up the Free World for its nickel and the ransom has been paid."

In the end, roughly one quarter of the necessary capital came from government agencies in the form of export credits or from lending institutions which receive most of their funding from the advanced capitalist countries. Much of this loan capital is classified as "aid" or "development assistance". But the following analysis of the Exmibal financing, which appeared in the *Northern Miner*, the house organ of the Canadian mining industry, suggests that it might more accurately be called corporate welfare: " . . . it would appear that there are certain advantages to obtaining funds from government agencies. First of all, the price is usually quite right . . . in comparison to commercial rates for money these days. . . . Furthermore, it seems logical that international involvement is a form of insurance against expropriation. . . . After all, a developing country would certainly think twice before simply confiscating an operation which was financed by countries from which it gets foreign aid. It is a bit of a deterrent against biting the hand that feeds, and all that . . .

"With the huge amount of capital now being required to develop an orebody of any size, there is no doubt that this formula will soon come into even more common usage."[16]

The current financing* of Exmibal includes the following loans:

    i. a $15 million loan from the IFC itself

    ii. a $13.5 million loan from the U.S. Export-Import Bank (EXIMBANK)

    iii. a $17.25 million loan from Canada's Export Development Corporation (EDC). (According to the EDC, the loan is to Exmibal, for "the Guatemalan borrower to purchase Canadian equipment and services for a nickel mining and smelting operation in the vicinity of Lake Izabal in Guatemala." The only reference to Inco by the EDC is to note that the Company's provision of technical services to the project.[17]

    iv. a $6 million loan from the Central American Bank for Economic Integration (CABEI). Like ICAITI, CABEI is part of the Central American Common Market concept. The Bank is a creature of U.S. policy, conceived by the United States Agency for International Development. Its purpose is to help attract private capital to the region with generous loans.[18])

    v. a loan of around $5 million from the Export Credits Guarantees Department of the United Kingdom

    vi. a loan of $9 million from the Orion Termbank

    vii. a $13.75 million loan from the National Westminister Bank of London, England

*The successful mobilization of capital coincided with the coup in Chile which removed the threat posed to private investment by Salvador Allende's government.

74 THE BIG NICKEL

        viii. a loan of $13.5 million from the Chase Manhattan Bank under guarantee from the U.S. EXIMBANK

        ix. equity capital from the project sponsors, Inco and Hanna Mining. (In 1976, Inco's projected capital expenditures were $500 million, the major portion of which was earmarked for overseas lateritic developments.[19])

That the Canadian state contributed to this project should not in itself cause surprise. The history of Canadian government support to private industry dates back to the construction of the CPR and Confederation. But it is instructive to note the net effect to the Canadian economy of this type of "foreign aid".

Theoretically, export credits provide jobs by stimulating the manufacture of export products. In fact, Inco has been given assistance in diversifying its base of operations and enlarging its opportunities to dictate the circumstances under which it will operate in different countries. The future of towns like Sudbury and Thompson may not depend on the extent of their ore bodies, but rather on where in its worldwide empire Inco can extract the greatest profit.

THE POPULAR OPPOSITION TO EXMIBAL

Political opposition in Guatemala has not been confined to the rural guerrillas of Zacapa and Izabal. One of the other forms it has taken has been an open campaign specifically against the concessions extended to Exmibal by the dictatorship.

In May, 1969 an ad hoc group of academics, unionists, and opposition political leaders convened a public inquiry at the National University of San Carlos into Exmibal's intrigues. This commission made the following recommendations:

        i. a new mining code should be devised to replace the one passed in 1965, the "Inco code" which was made to order for Exmibal;

        ii. Exmibal should be taxed directly so that a definite sum of money would accrue to the government for every ton of ore extracted; or, the state should ask other companies to bid for the mine or operate it directly, itself;

        iii. The decision to allow Exmibal to repatriate profits freely, in contradiction of normal foreign exchange regulations, should be reversed by the government;

        iv. Exmibal should be denied classification as a transformation industry.[20]

This Report was only the highlight of the rising tide of resistance to the Lake Izabal project. As Mendez Montenegro left the presidency early in 1970, a cloud of uncertainty hung over Exmibal.

The company was quick to launch a counter-offensive through the media and continued to cultivate its contacts among members of the ruling elite. And the inauguration of Colonel Arana, the "Butcher of Zacapa", in March, 1970 had to bode well for the project.

Arana almost immediately ordered a complete review of the concessions to Inco's subsidiary to be done by his Minister of the Economy, Gustavo Miron Porras. The latter just happened to be a former employee of Exmibal. Not surprisingly, his report was essentially a "whitewash". In 1971 the final agreement was ratified.

INCO'S SEARCH FOR OIL — THE BELIZE CONNECTION.
In 1970 the government of Belize, a tiny British Protectorate which adjoins Guatemala on the north about 25 miles from the Exmibal site, granted exploration permits to Ajax Oil and Ariel Petroleum. These Delaware-based companies are Inco and Hanna subsidiaries, respectively. In 1975, the government of Belize did an about-face and cancelled the permits on an area of some 1500 acres in the Toledo district. The concession covering the area was handed over to the Anglo Exploration Co. of Houston. Inco and Hanna initiated court action against both Anglo and the government of Belize.[21] Inco's quick response indicates that there may be enough oil involved to greatly enhance the profitability of Exmibal. The likelihood of this was underscored in August, 1976, when the entire Ajax Ariel concession was sold off to Esso Ventures Ltd., a unit of the giant Exxon corporation. Under this sales agreement Exxon was obliged to conduct extensive geophysical operations and drilling offshore in the first six months of 1977, just when Exmibal was scheduled to go on stream.[22]

Against this background Guatemala was doing some sabre-rattling, attempting to press its long-standing claim of sovereignty over Belize. Troops were massed along the Belize/Guatemala border and a special squad of commandos modeled after America's Green Berets was trained in the vicinity. These so-called "Purple Berets" were also part of Guatemala's threatening posture toward Belize.[23]

On the diplomatic front Guatemalan vice-president Mario Sandoval Alarcon was touring various world capitals looking for support for Guatemala's claim to Belize at the United Nations. In November, 1976, the U.N. Decolonisation Committee rejected Guatemala's claim.

Arana, however, was not content just to use manipulation to achieve his ends. Living up to his campaign promise to turn the entire country into a Zacapa, he imposed a state of seige in November, 1970. Though this suspension of normal legal rights was designed to allow for increased official repression, it also seemed to facilitate an increase in the activities of right-wing "death squads". The conclusion is inescapable that these terrorist groups, including Ojo por Ojo (Eye for an Eye), Buitre Justiciero (the Justice-giving Falcon), and Mano Blanco (White Hand) were working in collaboration with the authorities.*

Some of the most vocal critics of the Exmibal deal were victims of the terrorists during this state of seige, a time in which the government should have found it easier than usual to control terrorism. Alfonso Bauer Paiz, a law professor and member of the ad hoc commission, was shot by assailants but survived. In the same month of that attack another member of the commission, Julio Camey Herrera, was machine-gunned to death in his car. And two months later, just before the signing of the final Exmibal agreement, Adolfo Mijangos, the foremost critic of the project, was murdered.

> *The army believes that its mission is to pacify the country in the interests of U.S. security. The Colonels surrounding Arana have fallen in with the U.S. plan of using national armies to do the police work for them....*
>
> *Every time I leave home to go to my office, my wife wonders if it's the last time we'll see each other. One hopes for a quick death. That's all.*
>
> — Adolfo Mijangos, December 10, 1970[24]

Mijangos, also a law professor, had been one of the four opposition deputies elected when Arana came to power. A paraplegic confined to a wheelchair, he was shot as he was going from his office to his car.

LAKE IZABAL TODAY

In July, 1977, Exmibal was officially dedicated and by the end of that year was expected to be exporting its product. The extractive part of the operation is capital intensive strip mining. Power shovels scoop away the hills in 22 foot strips and 35 ton trucks haul the ore to the processing plant on the shores of Lake Izabal. After the ores are reduced and smelted, the nickel matte is shipped down the Lake and eventually to the port of Santo Tomas on the Caribbean, where it is loaded onto ships bound for those countries which refine nickel.

Once it is operating at full capacity, Exmibal will employ 900 people, including expatriate staff. It is the largest private sector investment in Central America.

With an unemployment rate of 20 percent and underemployment approaching 50 percent† in a total population of five million,

---

*This conclusion is re-inforced by the election in 1974 of Mario Sandoval Alarcon, founder of Mano Blanco, to the vice-presidency.

†"Underemployment" records those who are only able to find marginal occupations, seasonal or part-time, or in some way deficient in fully using the human potential of those so confined.

Guatemala will not experience any significant effect in this regard from the Lake Izabal complex. New problems will be introduced to the area, however.

For each ton of mineral extracted, over one ton of waste will be produced, presenting a difficult disposal problem. Strip mining itself is also environmentally devastating. But the ecological dangers do not stop at the mine site. Large amounts of fuel oil are being used in this plant, which has necessitated the building of a storage facility capable of holding 146,000 barrels at El Estor on the shores of Lake Izabal. The bunker oil for these tanks is being shuttled by two container barges from a loading point in the Bay of Amatique on the Caribbean through the Rio Dulce and the Golfete and across Lake Izabal, a total distance of 50 miles.

To avoid demurrage charges, there is also a depot close to the off-shore loading point. Because the Bay of Amatique is quite shallow, barges are being used to unload tankers anchored far off-shore, to transport the oil to the depot or tank park. From there, it is trans-shipped to the mine site. Needless to say, Exmibal claims that "no accidents in the Rio Dulce or Lake Izabal are anticipated, and the likelihood of spills at the offshore loading point has been minimized."[25]

With a revolver slung conspicuously on his hip, General Kjell Eugenio Laugerud, Guatemala's President, officially opened Inco's nickel mine in that country's El Estor province. The July 12, 1977 ceremony was kicked off with the hoisting of the Canadian and Guatemalan flags — Canada's being raised by Chargé d'Affaires William Taylor, head of the Canadian mission. The opening was heralded in the Guatemalan press with such headlines as "Guatemala: Nickel Capital of Central America."

The installation is seen by the Guatemalan government as a model of foreign investment. Not only does it open up a new export sector, but the project's Canadian face represents a diversification away from the traditionally high-profile American investment. Part of this progressive image was the President's call for humanitarian treatment of the workers: "It is necessary that the Exmibal Company recognize that what stays here is the sweat of the Guatemalans, who are men like whoever else and, that while they aren't blond like the North Americans, they are still human beings."

Two reports issued the same week as the Exmibal opening highlight an unofficial, though certainly popularly-held belief that the new mine represents little more than a giveaway, ultimately related to the government's repression of popular organizations and trade unions.

The first document, entitled *Exmibal Against Guatemala*, was published by the Economic Science Faculty of the University of San Carlos. This eighty-page report details the inadequacies of the agreement between Exmibal and the government, stating that national interests are overlooked and comparing this new investment to others which have gone before — notably the notorious United Fruit Company. It criticizes the fact that the company, not the government, will regulate the rate of resource extraction and control marketing and pricing. Concern is also expressed over the accounting procedures used by Exmibal and the threat to the ecology of Lake Izabal posed by the extraction procedures. Finally, it reiterates earlier criticisms of the taxation arrangements which exonerates Exmibal from 50 percent of the normal income tax in exchange for government participation of up to 30 percent.

The second report, *Fascism in Guatemala: A Vast Repressive Plan Against Popular and Trade Union Movements*, was released by the National Committee of Trade Union Unity (CNUS). CNUS calculates that with the rapid rise in prices over the last five years (77.4 percent) and the lack of corresponding increases in income, the salary of the average worker now fails to cover even *half* of the *minimum* dietary requirements for a family of five. With less than 3 percent of the Guatemalan working class organized, workers have been hard put to fight the decrease in their real incomes, let alone better their conditions.

This report also exposes an extensive, officially-sanctioned campaign of repression and murder which has been used to intimidate workers in their attempts to unionize. In 1976 alone there were 286 assassinations reported by the press. The targets of this repression are shown to be predominantly trade unionists, students, peasants, lawyers, and others participating in opposition movements. Just over a month before Exmibal opened Mario Lopez Larrave, a labour advisor and past dean of the Law Faculty at the University of San Carlos, was murdered by machine-gun fire in downtown Guatemala City. Larrave had been actively assisting CNUS. At the end of July the bodies of two student leaders, Leonel Caballeros and Robin Garcia, were uncovered. Credit for Garcia's death was publicly claimed by one of Guatemala's many paramilitary groups, which are linked by CNUS to counter-insurgency courses financed by U.S. aid for officers in the Guatemalan armed forces and police.

The workers at the Exmibal mine are not organized. Indeed, the form which prospective workers are given to fill out asks among other things if they have previously been trade unionists. Watching their pistol-packing president officiating at the mine's opening, Inco's Guatemalan workers probably had few doubts about the difficulties they will encounter in trying to establish a union there.

The main benefits from Exmibal seem likely to accrue to those corporate interests represented by Inco and to a lesser extent, Hanna. For the former, this development will help maintain its dominant position in the capitalist world's nickel industry. For both, profit seems assured. And the fabricators of specialty steels in the United States will continue to be assured of supplies of this strategic raw material.

The Guatemalans who are most likely to gain anything from this project are those few who have allied themselves directly to transnational corporations like Inco. The only Guatemalan on Exmibal's Board of Directors, Gabriel Biguria Sinibaldi, is a good example. A lawyer and financier, and key lobbyist for the Company, his family is one of the twenty most powerful in the country. It has interests in banking, commerce, and agriculture.

Perhaps those like Biguria Sinibaldi will gain even more influence in managing the country at the expense of those based in the more traditional agricultural export sector. Regardless, the state seems likely to be assisted in the development of a more efficient, better-equiped repressive apparatus.

For the vast majority of Guatemalans, Exmibal represents nothing if not the beginning of another cycle in a long history of foreign domination and exploitation.

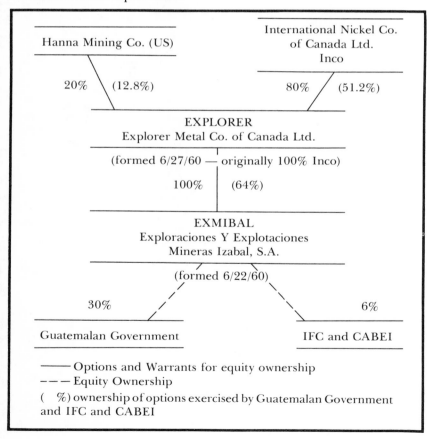

... And in Indonesia

In 1968 Inco was granted exclusive exploration rights to a 25,000 square mile tract of land on the island of Sulawesi, Indonesia. For more than fifty years, this island, one of the larger in that Southeast Asian archipelago, had been known to contain sizeable deposits of lateritic nickel.

Why had such promising reserves been left dormant for so long, particularly in the two decades after World War II when nickel seemed in such short supply? And why suddenly in 1967, did Inco find itself competing for this concession with ten other transnationals including the Japanese conglomerate Sumitomo and a French-American consortium of Le Nickel and Kaiser Steel?

In 1970, an Inco executive explained his Company's decision to pursue this prospect: "Indonesia, which has, in our estimation, vast resources, is also becoming quite stable politically... the basic concern is the ability to eventually repatriate earnings and some part of the investment made. There is just no point in investing large quantities of money in what you would consider unstable situations, where the chances are you might not get your money back, let alone any profit that might accrue from it. I think the most important consideration is the freedom of movement of funds, and the necessarily favourable climate to make investments satisfactory and relatively safe."[1]

Reading between the lines, it becomes obvious that here, as in Guatemala, the social and political climate was the prime determinant in the investment decision. To fully appreciate this situation, again an outline history of the host country is helpful.

THE QUEST FOR LIBERATION
As in Guatemala, so too in Indonesia the final years of World War II signalled a change in regimes. In the latter, it meant first the end of

the Japanese occupation of this sprawling island grouping and second, the culmination of a liberation struggle against the Netherlands, the pre-war colonial power.

In August, 1945, the Indonesians proclaimed their independent statehood and nationalist leader Achmed Sukarno became the new Republic's first president. Holland contested this proclamation and tried to re-assert its rule. There followed over four years of skirmishing, manoeuvering, and negotiating. Eventually the debate reached the Security Council of the United Nations. The Americans, eager to promote their own interests at the expense of the moribund European empires, supported the Indonesian nationalists.* Finally, the Dutch and the Sukarno regime reached an accommodation and Indonesia became fully sovereign on December 27, 1949.

The legacy of the independence struggle and an extremely complex social and cultural mix, combined to form a peculiarly ambiguous political situation. Sukarno and the intellectuals around him provided a loose ideology called Pantja Sila, the five principles of the unity of Indonesia (nationalism), humanism (internationalism), the sovereignty of the people (democracy), social justice, and belief in one God. The last principle exemplifies the vagueness of them all — it was meant to pacify the adherents of orthodox Islam, while avoiding the declaration of a Moslem state.

Sukarno himself enjoyed immense personal prestige to which most political factions tried to appeal. He attempted to maintain his popularity by balancing competing interests. Below the titular Father of Independence clashed the active competitors for power, the military and various political factions. The Republican Army, though not forged in a protracted liberation struggle, provided a centralized, nationwide bureaucracy which increasingly assumed administrative functions in state-run economic enterprises. The political parties, for the most part based on traditional religious or community relationships, were incapable of attaining a national perspective and rising above the patronage approach to government. One partial exception was the Communist Party of Indonesia (PKI).

Reorganized from oblivion after World War II, the PKI quickly abandoned any pretension of leading an armed struggle for revolution. Instead, it sought to broaden its base by embracing the vague but curiously militant nationalism of Sukarno, while at the same time subjecting the daily administration of government to an apparently

---

*The U.S. government may have been motivated in part by a fear of the possible consequences of a protracted armed struggle by Indonesian nationalists (of the sort that developed in Indochina).

radical criticism.

As the government was notoriously corrupt and inefficient, this strategy soon brought results, particularly as many Indonesians preferred to separate the revered Sukarno from the wrongdoings of his "evil" advisors and subordinates. In the first national elections in 1955, the PKI received over six million votes, 16.4 percent of the total. And by 1964 it had two and a half million party members, the largest of any non-governing Communist Party in the world, and it controlled mass organizations with 16 million members.

In 1956, Sukarno determined to take a more active role in government. Dissatisfied with the fractiousness of a multi-party democracy, he introduced a concept of Guided Democracy. With the collaboration of the Army, he dismantled the parliamentary system, nationalized Royal Dutch Shell and some smaller foreign interests, and instituted a moderate land reform. In the longer term these measures merely increased the opportunities for patronage and corruption because they were not accompanied by any overall strategy for the transformation of the social system in its entirety and did not allow for significant popular participation.

Around the same time, Sukarno, always the consummate politician, assumed a prominent role on the international "scene" by hosting the famous Bandung conference of the non-aligned countries. Increasingly, his attitude towards the imperialist powers became more militant. But at home, the economy deteriorated and military technocrats had to be installed in more crucial posts, both in government and in the state-run oil, tin, and agricultural enterprises.

Between 1953 and 1965, the country's dietary staple, rice, increased in price ten fold. The gap in living standards between the mass of poor peasants and the urban dwellers of the *kampongs* (shantytowns) on the one hand, and the elite, especially the government and military officials on the other, increasingly contradicted the traditional income distribution which had been characterized as "shared poverty". This polarization became increasingly visible as the new elite sought to hedge against the country's apparently uncontrollable inflation by acquiring large land holdings. With a steadily mounting foreign debt, Indonesia appeared headed for collapse, if not political upheaval. The first challenge to the fragile equilibrium came in 1958 with a full-fledged rebellion in the Outer Islands.

Java, the most populous of the Indonesian islands, had completely dominated the country from its birth. The resulting regional antagonisms were heightened by a difference in religious outlook between many of the Javanese, who are nominally Islamic, and the much stricter Islamic population of the other islands. The "modernists" among these latter were represented by the Masjumi, a Westward leaning, "free enterprise" party. With the support of the CIA, the Masjumi politicians and a number of disenchanted Army officers attempted to lead a break-away. Centred mainly on the islands of Sumatra and Sulawesi, this rebellion received air support from Civil Air Transport, a CIA-owned company. And U.S. Secretary of State John Foster Dulles gave his moral support, warning of the dangers of Communism in Indonesia. However, loyalist forces were able to quell the up-rising and Sukarno used the whole episode to legitimize his elimination of parliamentary democracy.

The failure of the Outer Islands rebellion forced the United States to revise its strategy. Not only had some American investments been nationalized, but the drift of the country under Sukarno's leadership seemed to be accelerating leftward. The prospect of permanently loosing Indonesia's wealth and investment opportunities became real.

If we lost Vietnam and Malausia, how would the Free World hold the rich empire of Indonesia?

— President Dwight Eisenhower, 1953[2]

The Americans adopted a now familiar plan — they focused their attention and their aid on the country's officer corps.* This process developed as follows: "High-ranking Indonesian officers had begun U.S. training programs by the mid-fifties. By 1965 some 4000 officers had been taught big-scale army command at Fort Leavenworth and counter-insurgency at Fort Bragg. Beginning in 1962, hundreds of visiting officers at Harvard and Syracuse were provided with the skills for maintaining a huge economic, as well as military, establishment, with training in everything from business administration and personnel management to air photography and shipping. (The U.S. Agency for International Development's) 'Public Safety Program' in the Philippines and Malaya trained and equipped the mobile brigades of the Indonesian military's fourth arm, the police."[3]

Similarly, Indonesian civilians were trained in American universities on Ford and Rockefeller Foundation scholarships. Properly selected and guided, graduates of Harvard, MIT, and Berkeley could

---

*This approach parallels the courting of the Chilean military which continued after the victory of the Allende forces in 1970 had prompted the U.S. to suspend all non-military aid to that country.

be expected to defend American interests upon their return home to influential positions in the bureaucracy.

American direct financial aid complemented this "educational strategy". By 1962 aid to Indonesia for domestic projects had been largely curtailed and by 1965 all forms of development assistance from the U.S. had been completely cut off. However, between 1962 and 1965 military aid totalled $35.8 million, compared with $29.7 million for the entire post independence period prior to that time (1949-1961).[4]

Communism is bound to win out in Southeast Asia . . . unless effective countervailing power is found in some groups who have sufficient organizational strength, goal-direction, leadership, and discipline . . . those best equipped are members of the national officer corps as individuals and the national armies as organizational structures.

— Guy Pauker, Rand Corp. consultant
closely associated with U.S. policy formation on Indonesia[5]

In the meantime, the PKI had embarked on its first independent initiative in the greatly ignored countryside of the nation. Using some moderate land reform laws passed in 1959 and 1960, the Party's cadres began to work among the peasants, informing them of their rights and encouraging them to assert themselves. By 1963 this work was beginning to show results and there were indications of a developing class consciousness in the rural areas.

As the peasants began to demand an end to unproductive land speculation and their legal return in share-cropping arrangements, the ruling circles in Jakarta, the capital, began to abandon their tolerance for the PKI. The latter, nevertheless, continued to restrain its urban trade union followers, despite a disastrous inflation.* Its official line was to push for NASAKOM, Sukarno's plan to incorporate nationalism, religion, and communism, as represented by the major parties, into a government of national unity.

But with Sukarno ill, the jockeying for the succession had begun, and the Foreign Minister Subandrio† seemed well-placed. Early in 1965 he had helped conclude an alliance with the People's Republic of China and he had skillfully developed the anti-imperialist foreign policy initiated by Sukarno. It began to look as though the mantle of the

---

*In late 1964, the official exchange rate was 516 rupiah to the U.S. dollar, while the black market rate was 4000 to one; three months later the street rate was 12,000 to one with no change in the official price.
†Many Indonesians are known by one name only.

Father of Independence was to be bestowed on the "left-wing" of the political establishment, including the PKI.

> *Perhaps overnight the General Staff or some younger members of the Officer Corps in Indonesia will strike, sweep their house clean, and rededicate themselves to higher purposes.*
>
> — Guy Pauker[6]

On September 30, 1965 a group of Javanese officers led by Colonel Untung of the Presidential Palace guard, seized and immediately executed Army Chief of Staff Achmad Yani and five other generals. They claimed to be forestalling a coup attempt by the General Staff. Though Untung's forces were able to occupy some of the important installations around the capital city, their success was short-lived. One of the remaining officers capable of rallying troops against this "pre-emptive coup", General Suharto, barely escaped capture. Within a day, he and other right-wing officers, notably General Nasution, were able to organize a successful counter-coup.

Though the PKI was not officially involved in Colonel Untung's plan, one of its leaders, D.N. Aidit, and others at a lower level in the Party may have played some role in it. Regardless, Suharto was able to turn the momentum of events to his advantage, justifying his seizure of power as being necessitated by the "impending Communist takeover".

The military quickly consolidated its position, then moved to eliminate the only force in the country with the potential to mount any sustained opposition to it — the PKI.* Most of the Party's leaders were summarily executed. Then, in mid-October, the Army unleashed one of the more bloody rampages of modern history. By example, bands of civilians were encouraged to slaughter anyone suspected of affiliation with, or sympathy for the Communist Party. Some were motivated by religious zeal, some by right-wing ideological hatred, others by communal jealousies. Though this onslaught was intended for activists in the peasant leagues and the trade unions as well as those in the PKI, it quickly extended to whole families and to others with only tenuous associations with the ostensible targets.

> *The killings have been on such a scale that the disposal of the corpses has created a serious sanitation problem in East Java and Northern Sumatra where the humid air bears the reek of decaying flesh. Travellers from these areas tell of small rivers and streams that have been totally clogged with bodies. . . . River transportation has at places been seriously impeded.*
>
> — *Time* Magazine[7]

---

*This was a potential, not a capability. The PKI membership had been nur-

Estimates of the number of people killed range from 500,000 to one million. Tens of thousands were confined to special prison camps where they languish to this day. According to the human rights group, Amnesty International, the Indonesian government currently holds 55,000 political prisoners.* Very few of them face specific charges. In 1971, Attorney-General of the regime, Sugih Arto, explained his government's view of the matter: "We know for certain they are traitors, that they are ideologically conscious, but there is not enough evidence to bring them before a court. . . . It is impossible to say exactly how many political prisoners there are. It is a floating rate, like the Japanese yen vis-a-vis the dollar."[8]

Siti Suratih . . . is one of more than 55,000 prisoners of conscience who have been detained for ten years without trial in Indonesia. A trained nurse who worked fulltime and also managed to bring up her four children, her only conceivable crime could be that she married a member of the Communist Party. Her husband died in 1968. Yet Ms. Suratih is still in a concentration camp, awaiting a trial that may never come.
— Amnesty International: USA Letter #692, January, 1976

This "clean sweep" by Suharto and the military was precisely what American foreign policy planners needed to tip the regional balance of power in favour of the U.S. empire. Communist insurgencies in the Philippines and Malaysia had already been successfully repressed and the battle for Indochina was at that time escalating.†

Though the Americans were to lose Vietnam, Cambodia, and Laos, Indonesia remains as a bulwark against Communism in Asia. More to the point, it is part of a strategic bloc, what U.S. Presidential advisor Zbigniew Brzezinski calls a "Pacific maritime triangle". The political, economic, and security interests of this triangle, Japan, Australia, and Indonesia, are now inextricably linked to the maintenance of the American international system.

THE NEW ORDER
Soon after President Sukarno was formally deposed in March, 1967,

---

tured with a very moderate practice and was largely dependent on its leadership. The Party's underground structure, if it existed at all, was useless.

*The AI Report, 1976-77, cites new evidence indicating that the total number of detainees is closer to 100,000. It also notes that torture continues for many.

†Neither the U.S.S.R. nor China was particularly vociferous in its protest of this massacre, the former presumably because the PKI had sided with Peking in its dispute with Moscow, the latter perhaps because of the internal Chinese Party problems that were about to emerge as the Cultural Revolution.

General Suharto proclaimed a "New Order" (Orde Baru) for Indonesia. While maintaining the Pantja Sila earlier espoused by Sukarno, the New Order began to construct a new political framework to legitimize Suharto's power. But the most substantial aspect of this new approach was the opening of the country to foreign investment.

A "Stabilization Plan" had been adopted in October of 1966 and a new Law of Foreign Capital Investment had followed it. Tax holidays, provisions for unrestricted repatriation of profits, exemptions from import duties, and cutbacks in government participation in the economy were all included to attract foreign investors. The country's creditors looked upon this initiative with favour, in December, 1966 granting a moratorium until 1971 on all repayments of interest and long-term debts incurred prior to the coup.

> The U.S. can influence the outcome in Indonesia. This is a situation where our real and total strength can be brought to bear — not the power of guns and bombs, but the accumulation of wisdom, experience, talent for organization and capital wealth. U.S. business can participate in the shaping of history while serving its own self-interest — by helping to unlock Indonesia's rich resources of oil and minerals and timber, as well as to develop among its millions a market tuned in to twentieth century technology and productivity. Some major American corporations have already begun to respond to the dramatic change in the Indonesian environment.
>
> — *Fortune* Magazine, June, 1968

Foreign capital quickly gravitated to the lucrative opportunities promised, particularly in the resource extraction sector. In the years between 1967 and 1974, foreign investment in Indonesia totalled $3.89 billion.

These private investments have been complemented by huge injections of aid from foreign governments. In 1967, after agreeing to postpone Indonesia's debt repayments, the creditor-countries formed an aid consortium called the Inter-Governmental Group on Indonesia (IGGI). Much of the $4.5 billion disbursed as loans and export credits by the IGGI up until 1974 facilitated the operations of the transnationals, again those in the extractive sector in particular, by financing infrastructure — roads, bridges, ports, transportation, and communications systems.*

---

*Among the transnationals which are operating in Indonesia are Caltex (a subsidiary of Standard Oil of California and Texaco), Stanvac (another American joint venture) and Shell, all in petroleum; Freeport Copper and Rio Tinto Zinc in mining; and Imperial Chemical Industries and Bata Shoes.

With its 100 million people and its three thousand mile arc of islands containing the region's richest hoard of natural resources, Indonesia constitutes the greatest prize in the Southeast Asian area.
— Richard Nixon, 1967[9]

Suharto's gauge for measuring development under the New Order is the "rate of growth" of the economy. As one Canadian government document noted: "The Suharto administration is primarily concerned with maintaining a politically stable environment in order to promote maximum economic growth."[10] And using the crude indices so popular with economists, there has been "growth" under the Suharto regime.

But the benefits of such growth for most of the 140 million Indonesians are illusory. One study done in a good growth year, 1972-1973, indicated an actual decrease by one-third in real wages being paid in 20 Javanese villages.[11] According to the standards devised for Asia by the Food and Agriculture Organization and the World Health Organization, close to two-thirds of the population still have an inadequate diet. Average annual income is $150 per person.

Every bit as deceptive is the dynamic behind the country's economic "progress". One critic has explained: "The Suharto regime and its supporters talk constantly of the economic miracle which has brought Indonesia out of the Sukarnoist dark ages. But on closer examination, the real miracles appear to be a result of accounting, public relations, and the ever-mounting annual IGGI credits, rather than of any substantial reforms. ... It works this way: the IGGI credits finance hundreds of millions of dollars of imports each year which Indonesia would not be able to pay for with its foreign exchange earnings. The country's foreign payments are thus "balanced" by foreign loans, and the debts continue to mount."[12]

The external debt from the Sukarno years was $2.4 billion. According to the World Bank, the country's external public debt at the end of 1975 was $11.3 billion.

Not the least of the inducements to invest in Indonesia of the New Order is the supply of cheap labour. One American investment guide, *Doing Business in Indonesia*, claims that wage rates are even lower than in most other Asian nations.* The government boasts that workers are

---

*In *Population and Poverty in Rural Java: Some Economic Arithmetic from Sriharjo* (Ithaca, 1973), D.H. Penny and M. Singarimbun claim that the daily wage rate for agricultural labour would require saving for 450 years to buy 2.5 acres of riceland, compared to a period of "only" 150 to 200 years to accomplish the same feat in the extreme poverty of Kwantung, China in the 1920's.

now "caught up in the spirit of development". This "co-operativeness" of the workforce is related to the fate of the union movement under Suharto.

By 1965 somewhere between 20 percent and 40 percent of the wage earners had been organized into unions. The most powerful labour centre, SOBSI, was generally allied to the PKI. It was based in agriculture, mining, transportation, and communications. During the fall of 1965 its most prominent leaders and activists were killed or jailed. In May, 1966, SOBSI was officially banned and its 3.5 million members were denied the certificate of non-involvement in the Untung coup which had then become necessary to obtain employment.

Although strikes had been made illegal in "vital industries" with Sukarno's proclamation of martial law in 1957 and there had also been a practical ban on them in all large enterprises, the competitive union centrals had been able to exert influence on behalf of their members through political leverage.* Since the military takeover, that avenue of redress has been eliminated and the number of strikes has even further declined. Yet squeezed by rising prices and falling real wages, workers still resort to strike action on rare occasions. During one such strike, at a Japanese-owned textile plant in 1972, a representative of the Department of Manpower maintained that "(l)abour should guarantee the sound sphere for foreign investment, otherwise it would hamper the national interest and the economic growth."[13]

This "national interest" rhetoric has been a cornerstone of the military government's ideology. "National unity" and the "organic whole" of Indonesian society have been emphasized; any basic conflict between the interests of the elite and the lower classes is denied.

In order to systematize this corporatist ideology, the dictatorship has created its own "unions". One such organization is KORPRI, established by decree in 1971 when Suharto decided to force all government employees and workers in state-owned industries into one single union. Effectively it is one giant company union — the executive of KORPRI consists of top government officials. Similarly, the All Indonesian Labour Federation was created in 1973 to centralize control of the nation's workforce. Its general secretary, appointed by the state, is an officer of the OPSOS, which is part of the military's vast intelligence and security network.[14]

THE OLD DEPENDENCY IN A NEW FORM

The New Order is in content only a return to the pattern of foreign

---

*All the unions were associated with some political force — nationalist, Islamic, Communist, some even with the military.

control over Indonesian resources first introduced by the Dutch in the seventeenth century. Sukarno's era was merely a partial respite, in no way thorough and in no way capable of reforming the economic structure systematically constructed over three centuries.

When the Dutch established a trading monopoly known as the Dutch East India Company in 1602, the destruction of the self-reliant economy of these islands got well underway.* The indigenous peasants were forced to pay taxes in the form of cash or produce. That requirement forced them to cultivate the crops which the colonial traders were willing to buy.

In the nineteenth century, the Netherlands formally incorporated into its Empire the many islands which comprise Indonesia. In the process they extended their penetration of the indigenous economy by introducing laws which enabled private interests to take over some of the most fertile land in Asia for the cultivation of export crops. The most important of these crops was sugar. Other resources in which the Dutch invested included rubber and coffee, and tin and petroleum.

As their economy was transformed into one dependent on the needs of European markets, the people of Indonesia became increasingly impoverished. Not only was their agriculture moulded to service international trade, but local industries, particularly textiles, were actively undermined by the Dutch administration in order to perserve the colony as a market for goods produced in Europe.

As have all colonized peoples so too did the Indonesians resist this foreign intervention. In part through the struggle against the Japanese occupation during World War II and in part with some last ditch aid from the Japanese as a rearguard action against the Allies in the dying days of the War, the nationalist movement finally emerged in a powerful position to challenge colonial rule. This was but one of the independence struggles which gathered momentum in the immediate post-War period. The liberation movements of the Indian subcontinent, of China, Malaya, and the Philippines reinforced each other in the fight against colonialism.

The Americans expedited the final transfer of state power to the Indonesian nationalists in 1949 by threatening to cut off Marshall Plan aid to the war-shattered Dutch economy. The U.S. had hoped to facilitate the smooth realignment of Indonesia from one economic orbit centred around Holland to another centred around America.

In the short term, the American strategists did avoid the radicalization which occurred in the independence movements of China and

---

*This venture was similar in many ways to the Hudson's Bay Company set up by the British for Canada.

Vietnam. In the middle term, they faced failure, though the interests they represented sustained no substantial losses during the Sukarno presidency. In the longer term, they seem to have won, further penetrating the culture and political life of the subject nation in order to guarantee American economic dominance.

INCO AND THE IMPERIAL PRESENT

The political climate for investment in Indonesia clearly impressed the managers of International Nickel in 1967. But another factor making the Sulawesi deposits attractive was their proximity to Japan, one of the world's most industrialized countries with an annual nickel consumption of several hundred million pounds.

Historically, most of Japan's nickel has come from Le Nickel's mines in New Caledonia, though in recent years it has moved to diversify its sources of supply. By 1972 Japan was already importing a quarter of its nickel from Indonesia. It is this lucrative market which Inco seeks to corner by developing at Soroako what will rank as "one of the world's principal sources of (nickel)".[15]

After outbidding its competitors for the development rights from the Suharto government, Inco carefully constructed an alliance with the Japanese steel industry. Six Japanese companies* contributed 20 percent of the financing for P.T. International Nickel Indonesia, in return for a marketing agreement under which three of the companies will take most of the production from the first stage of the project for 15 years.

*Inco's most important Japanese connection is with the Sumitomo group, the third largest of the *zaibatsu*, the giant conglomerates which dominate that country's business. Sumitomo Metal Mining is Japan's number one producer of nickel.

Association with Mitsui, the second largest of the *zaibatsu*, gives Inco another connection with the commanding heights of the Japanese economy. A third connection is with the fifth largest trading company, Nissho-Iwai. Affiliated with two other *zaibatsu*, Sanwa and Dai-Ichi Kanyo, its main business is metals. It owns 44 percent of Akashi-Gokin, a manufacturer of nickel alloys.

Inco also has a direct equity interest in three Japanese companies. It has a 50 percent interest in Daido Special Alloys, the second largest producer of specialty steels in Japan. Its partner in this venture is Daido Steel Company, an affiliate of Nippon Steel, the capitalist world's largest steelmaker.

Tokyo Nickel is 40 percent owned by Inco. It has a nickel processing plant in Japan and produces refined nickel products. Shimura Kako, an affiliate of Tokyo Nickel, is 30 percent owned by Inco. It is an integrated manufacturer of nickel and owns a refinery which produces nickel and ferronickel. Thus the Company seems to have both direct and indirect links in Japanese heavy industry.[16]

As Inco holds a substantial interest in some of these companies, it will be selling part of its Indonesian production to itself, providing the opportunity for it to declare its profit at whichever point in the production process is most lucrative.* Both here and in Guatemala, Inco's size, access to financial and technical resources, and mobility, give it important advantages when it comes to calculating the benefits which will accrue to the host nation and those which will go to itself.

The capitalist world's largest nickel producer seems to be providing a bridge to two points of the "Pacific maritime triangle". Reducing that "triangle" to its economic reality, Inco is attempting to secure a foothold with the world's most dynamic steel industry. And in Japan, the U.S.A.'s staunchest political ally,† "Steel is the state".[17]

FINANCING P.T. INTERNATIONAL NICKEL:
CORPORATE WELFARE STRIKES AGAIN
On March 31, 1977, President Suharto dedicated the Soroako project, unveiling a plaque to commemorate the occasion. This ceremony marked the completion of stage one in a two part plan and signalled the start up of an operation with a production capacity of 35 million pounds. The first shipments of nickel matte were made in the summer of 1977 to Japan.

The second phase of the project, which should be completed in 1978, will add an additional 65 million pound annual capacity to the complex. In total the P.T. International Nickel Indonesia investment should be $850 million. As with most projects of a similar size and nature, the cost of it has escalated since the earliest estimates were made. But at least the production capacity has grown, too.‡ Worldwide inflation has been one factor in the cost increase. However, the dramatic increases in petroleum prices which have occurred starting in 1973 have had a particularly serious effect on the economics of the refining of lateritic ores. Because these ores cannot be concentrated before smelting, they require an energy intensive process to separate the nickel content. The use of oil as the main energy input, as has been the practice, reduces the profitability of lateritic projects.

*It is very difficult to assess the real importance of this sort of equity interest in a purchasing company. In any case, the Indonesian subsidiary is so established that two-thirds of the full production after stage two is "on stream", is going to be purchased by the parent company, Inco. Both arrangements seem to allow similar abuse.
†Though Japan is allied to the American world security system, it remains fiercely competitive with U.S. capital on the economic level. The steel industry is an excellent example of this competition.
‡In Guatemala, the cost rose while the productive capacity was reduced.

Apparently in response to this "oil crisis", Inco decided to construct a 165 megawatt hydroelectric generating plant on the Larona River close to Soroako. But this additional investment could only be justified by a larger project than originally planned. The result is stage two and a capacity three times the amount to be absorbed by the minority Japanese investors.

In spite of these steadily climbing costs, Inco has had little difficulty in raising the necessary capital. One third of the financing is by equity capital, of which the parent Company retains 96 percent. The remaining equity is held by the Japanese companies who provided a buyers' credit of $36 million, though they have opted to forego their equity in stage two. Eventually they will retain only 3 percent of the shares. There is also a provision for the eventual sale of 20 percent of the subsidiary's shares to Indonesians, at the rate of 2 percent a year starting after the entire complex is in full production.

Three international banking syndicates were arranged to raise $365 million of the remaining two-thirds of the investment to be financed by long-term debt. The first phase syndicate was managed by the Bank of Montreal, while the larger of the second phase consortia was managed by the Citicorp International Bank of London. The latter is owned by Citicorp, a holding company for the second largest bank in the United States, the First National City Bank. Included in this syndicate were the Bank of Montreal, the Toronto Dominion, the BNS International (Hong Kong) — a unit of the Bank of Nova Scotia, Morgan Guaranty Trust, Crocker National Bank, Chemical Bank of New York, Bankers' Trust Company, and the Asia Pacific Capital Corporation.[18]

But not to be forgotten among these lists of private lenders is the role of governments in capitalizing P.T. International Nickel. The Canadian Export Development Corporation (EDC) lent $17.25 million for the first phase of the project, $40 million for the second. The American EXIMBANK provided $13.5 million for the first, $35 million for the second stage. Government "export credit agencies" in Australia, Norway, Great Britain, and Japan were also able to rationalize assistance to Inco in its Indonesian endeavour.

Direct government loans and export credits are again only part of the overall picture. Though indirect public subsidies to corporations have been an integral part of the "free enterprise" system for years, even some mining executives seem surprised by recent tendencies: "By means of tax breaks, depletion allowances, fast write-offs and in some cases, even direct subsidies, government has always been an indirect contributor to the mining industry ... (recently) there has been a growing tendency for governments to devote funds to mining development — often not in their own territories or through companies registered in their own jurisdiction — but in foreign countries, and especially those falling into the category of developing nations. Sometimes this is done through the auspices of international bodies, such as the World Bank . . . or the International Finance Corporation. Often it is done through national bodies such as EXIMBANK in the U.S. or CIDA in Canada."[19]

The Canadian International Development Agency (CIDA) has recently increased its direct aid to Indonesia, paralleling the increases in "Canadian" private investment. In 1971 Canadian aid to this nation of 140 million people was less than $4 million. That year Prime Minister Pierre Trudeau visited Jakarta and the following four years CIDA aid rose fivefold.[20] In July, 1975, General Suharto visited Ottawa where he successfully concluded a $200 million "package deal" of aid, export credits, and loans.* The aid contribution placed Canada in fifth spot among the "donor-nations" of the IGGI and prompted Trudeau to boast later that year that Indonesia had become one of the top recipients of Canadian largesse.[21]

But what is the nature of these assistance programmes? CIDA has funded surveys of resources and infrastructural needs on several islands of the archipelago, including Sulawesi. These surveys, for the most part carried out by Canadian engineering firms and universities, are part of a masterplan for regional development which the Suharto government has devised with foreign assistance. The transnationals are to be the agents of that development.[22]

Aid projects are becoming so numerous in Indonesia that many Canadian companies are cashing in on their special areas of knowledge....

... firms like Bow Valley and Foremost International Industries Ltd. of Calgary, Westinghouse Canada Ltd. of Hamilton, MacMillan Bloedel, Sandwell Company of Vancouver and other oil, mining, and consulting firms are selling technology, services, and equipment to meet the needs of the country's rising industries.

— Toronto *Globe and Mail*, August 21, 1975

A CIDA loan has enabled Merpati Nusantara Airlines, a government-owned line, to buy eleven Twin Otter airplanes from deHavilland of Canada. These short take-off and landing craft will improve communications with the Outer Islands, in CIDA's terms, providing them with "pioneer air service similar to the bush services

*Twenty-five million dollars came from CIDA, $75 million from the banks and $100 million from the EDC.

that contributed to northern Canada's development in the 1920's and 1930's."[23] This analogy between the type of development which will be facilitated by this service and that which occurred in the Canadian North is appropriate. The average Indonesian will rarely make use of air transport, but the prospectors, geologists, surveyors, engineers, and businessmen interested in the resources and trade of these relatively isolated regions certainly will.

Not only is Canadian aid largely tied to private investment, it is also used to promote Canada's export trade. From 1973 to 1974 Canadian exports to Indonesia tripled and were expected to continue rising.[24] Foreign aid, trade, and private investment clearly work in tandem.*

INCO AND INDONESIA: THE BALANCE SHEET

According to Philip C. Jessup, Jr., managing director of P.T. International Nickel Indonesia, the Soroako complex will employ 4000 people when it is in full production in 1978. At present, around 2500 are employed.

"We also see major stimulation to more development in the future as a result of the economic spin-off effected from the new jobs created by P.T. Inco and the demands for services, food and other goods generated by our presence in this part of Sulawesi," he told the 80 dignitaries gathered at the end of March in 1977 for the official dedication of the mine site.[25]

All mining, including open-pit and strip mining, is highly capital intensive. Most of the initial investment is for the purchase of sophisticated mining and processing machinery and techniques and the construction of transportation facilities. The disbursers of government aid in Canada and the other "benefactors" of Indonesia all know that nearly all of these purchases are made in the industrialized nations. As the Department of Industry, Trade and Commerce noted, the Inco project presented "an excellent opportunity for Canadian sales of a wide range of equipment, from safety supplies to mining equipment."[26]

For hundreds of years Buginese ships have sailed the seven seas from Sulawesi ports, carrying the island's products to market. Today, in the mountainous interior of this lush, green island, the space age metal nickel is being mined and processed for shipment to market, contributing to the progress of the world and to the island's economy."[27]

— Inco publicity pamphlet

---

*The target for aid set by the Pearson Report, one percent of the GNP, must be considered in the light of Canadian business needs, not moral obligations or humanitarian ideals.

The actual design and construction of the mining complex has been supervised by Bechtel Ltd., the San Francisco-based engineering giant, and Monenco, a Montreal-based transnational. As many as 11,000 people have been employed temporarily in construction and development.

It is the intention of Inco management to retire its Indonesian branch's debt in the first eight years of operation. Though its production should gross $200 million a year once the second stage is functioning in 1978, revenues to the Indonesian state will probably be around $8 million annually for the first decade of operations. For the remaining 20 years of the contract, taxes will be increased by an additional $20 to $30 million yearly.[28]

At the end of the 30 year work contract which Inco holds, its subsidiary should have sustained sales of $6 billion at current prices.* At the most, the Indonesian nation will have received $840 million in revenues from the extraction of its non-renewable resources.

The real tragedy is not that such a widely touted investment, generously assisted by governments in a number of countries, will employ only 4000 people in a nation which needs work for 15 million. Nor is it that the national revenues from this project after 30 years will barely be equal to the initial capital cost of the Soroako complex. Rather it is that this 30 years of revenue in total will not have been sufficient to service† Indonesia's debt for any one of those 30 years.

To keep their country attractive to Western investment, Indonesia's military rulers are plunging it further into debt on an ever-widening spiral. In the process, they have refined a power system based on corruption and repression, a system which feeds itself and grows out of control. In an interview in 1975, Carmel Budiarjo, a former political prisoner, tried to explain the extent of this corruption: "At every stage of any operation, be it government, private enterprise or foreign private enterprise you certainly have to put out a tremendous amount in corruption in order to get approval for the use of land, approval for the use of personnel, approval for the licensing of whatever you have to do. You have to pay far and away above the official rate for whatever these services may be."[29]

In 1976, Philip C. Jessup, P.T. Inco's managing director began to get a new perspective on the military dictatorship which had made its country safe for companies like his own. Jessup attempted to dismiss from his employ a lawyer from one of Jakarta's top law firms. The result was first a civil libel suit, then a charge of criminal libel. Suddenly

---

*This work contract can be renewed.
†That is, to repay the interest and retire on schedule part of the principal.

he found himself facing a four year prison term, despite the lack of foundation for the indictment.

When the case got underway, the presiding judge contacted Inco's lawyers, inviting their client to pay a bribe. Fearful of the complications such collaboration might have, both in Indonesia and in the United States where investigations of corporate bribery were underway, Inco's head office refused the invitation. Instead the ambassadors of the U.S., Canada and Australia were asked to intervene, which they tried, unsuccessfully. Finally, an even higher level intervention was arranged from New York and on the day sentence was to be pronounced, the case was dismissed.[30]

What of the living conditions of the average Indonesians? Three years after his assumption of the Presidency, General Suharto addressed their plight: "Only a part of us — just the smaller part — have enjoyed the fruits of independence. The greater part, however — soldiers, civil servants, small peasants, and labourers — still lead a difficult life. They still live in inadequate houses, in alleys and dirt slums without fresh drinking water and without electricity, with a low degree of nutrition, deeply concerned about the schooling and education of their children, as if looking forward to a dark future. Why, after being independent for a quarter of a century, are we still living under such conditions?"[31]

Is there reason to believe that Suharto's description has become outdated, that his regime has started to look for real answers? Not according to International Policy Report, a Washington-based publication: "Tight control is maintained on the press; approved papers enjoy advertising and circulation support from the government. Public debate is all but impossible on any issue of political importance. The workers' rights to organize are of course curtailed. Even as the oil income misleadingly increases the per capita annual income from about $100 to $150, the gap between the poor and the rich grows even wider."[32]

It is ironic that Philip C. Jessup, father of the managing director of P.T. International Nickel Indonesia, is on the Board of Advisors of the Centre for International Policy, the publishers of this highly critical report. In 1949, Jessup Sr., as American ambassador to the United Nations, forcefully presented his country's case for an independent Indonesia.

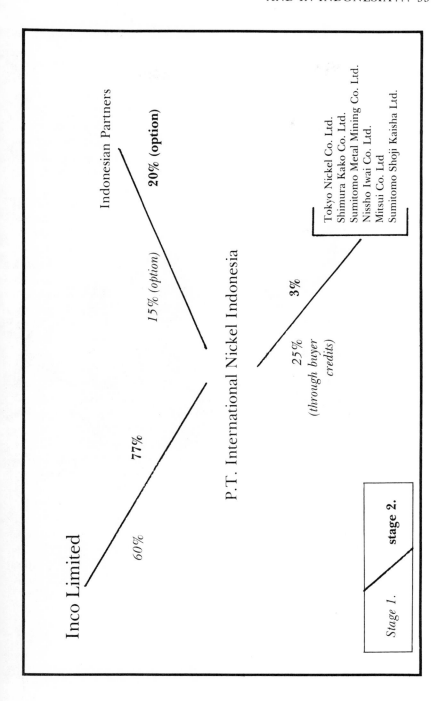

New Growth Strategies: The End of Motherhood

In 1976 the International Nickel Company of Canada became Inco Ltd., thereby institutionalizing the name which the Company had familiarly borne for years. Edward Grubb, chairman of the board, had noted that Canadian subsidiaries of foreign corporations often simply add "of Canada" to the parent company's name.* Consequently, the old name suggested that Inco was not the parent company but a mere branch plant operation.

It would seem that this superficial change reflected a concern for nationalist sentiment in Canada — a sentiment earlier managers of the corporation had often been careless enough to ignore. But more important, it was a reflection of a new, more aggressive growth strategy ending what Grubb termed the "motherhood approach to marketing".

This new strategy has not only affected all aspects of the Company's nickel operations, but has also led to a diversification outside of its normal field. Again, in this respect, the dropping of "nickel" from the Company's name is appropriate.

During the first two decades of Inco's existence (1902-1922), military demand for nickel was the main spur to its corporate growth. The depression which followed World War I severely affected the Company's sales and quickly prompted a vigorous research and development programme to promote new consumer uses for nickel.

This new marketing drive began just prior to Inco's merger with the Mond Company which gave the Company a near total monopoly in nickel. So, throughout the remaining years of the Twenties, Thirties, and most of the Forties, it was enough to sell the benefits of nickel in general, for that almost automatically meant selling Inco nickel.

The rise of Falconbridge and the other smaller producers during

---

*For example, General Motors of Canada, or even better, American Can of Canada.

the Fifties reduced Inco's share of the market. But this was a rapidly expanding market which the Company seemed ill-prepared to serve. According to *Fortune* Magazine: "The chronic and intense character of the nickel shortage suggests that Inco, despite its growth, has been overly cautious.... (I)n 1950 Inco was caught short by a peacetime upsurge in industrial demand for nickel. Nickel has been grieviously short ever since. Says a Canadian competitor, 'They did such a good job of building the market that they never noticed when it went right by them'."[1]

It was no longer enough to create markets for nickel in general when other companies were also supplying the product.

The American war in Vietnam brought a new surge in nickel demand in the Sixties. What should have been "salad days" for Inco turned out to be a period of only moderate profitability. Even with its new mines in Northern Manitoba, the Company could not keep up with demand. In 1966, Inco's management embarked on a multi-million dollar programme to increase productive capacity by one third. In 1969, as these changes were being implemented, a bitter four month strike in Sudbury dealt the Company a severe blow. The settlement was a clear victory for the Steelworkers' Local 6500. Still, Inco's troubles were not over.

The recession of 1970-71 resulted in a 25 percent drop in sales and a 55 percent decline in profits to "only" $94 million in 1971.* Stock prices tumbled to $24 a share from the previous year's high of $48.

In 1972 a new management team under Edward Grubb took over, with a mandate to cut costs, modernize, and rationalize operations. Grubb had the reputation of being a ruthless cost-cutter. At a previous posting as head of one of Inco's British subsidiaries, he eliminated the jobs of one fifth of his employees.[2]

Canadian workers soon felt the Grubb touch. In the summer of 1971, the Steelworkers' Sudbury bargaining unit had 18,224 members. Two years later, the figure was 12,522.[3] Jobs were eliminated by attrition, some workers were forced into early retirement and others were simply laid off.

But cutting payrolls was not even half the task. In an attempt to tie major customers more firmly to Inco, the Company started signing them to binding contracts. In the 1971 slump most of the drop in nickel demand had been absorbed by Inco. Grubb wanted to end that status of being what he called "the supplier of last resort."

By the end of 1974, the new approach seemed to be paying off, with earnings nearly hitting $300 million. The following year, despite a downturn in demand related to general recession, the Company main-

---

*Inco paid no federal taxes that year.

tained satisfactory profit levels. Before his retirement in 1977, Grubb expressed some concern about Inco's profits in 1975 and 1976. "But looking at other producers, we've done fairly well — and we've never been in serious trouble," he admitted.[4]

A good part of this new strength in riding out cyclical demand for metal products results from the third pillar of Grubb's management strategy — diversification. A corporate task force decided that the acquisition of a company well-established in energy production would be desirable. Considering the increasing demand for pollution-free power sources, the committee settled on the battery industry and one particular firm, ESB Ltd.

Formerly known as the Electric Storage Battery Co., this Philadelphia-based transnational is the largest manufacturer of packaged power sources in the United States. ESB produces such familiar brand names as Ray-O-Vac, Exide, and Atlas, as well as batteries for commercial and industrial application. It has 96 plants, nine of them in Canada and others in South Africa, Brazil, Guatemala, and 14 other countries.

Inco's acquisition bid in 1974 was resisted by ESB management which felt that the offer of $28 for each ESB share was not a real reflection of the company's value. It labelled Inco's move "hostile" and solicited a counter-offer from United Aircraft Corporation (now United Technologies), a company which had expressed interest in ESB in the past. ESB went so far as to invoke the spectre of "foreign ownership", a strange twist given Inco's own history. After some competitive bidding, an Inco offer managed by Morgan, Stanley and Company was accepted. The final price was $41 per share, making the total cost to Inco $234 million.

In January, 1976 the U.S. Justice Department initiated civil proceedings against the Company to force it to divest itself of ESB. The charge is that the acquisition is an "anti-competitive practice". Inco is "denying the material allegations of the complaint", according to its 1976 Annual Report, and both "parties are engaged in pre-trial discovery proceedings."

A second acquisition was the purchase in 1975 of Daniel Doncaster and Sons, a British fabricator of precision machinery which produces turbine blades, gear transmission systems, hydraulic systems, and a wide range of specialty steels. It is an important supplier to the British aircraft industry. As Doncaster is supplied with nickel alloys by Henry Wiggin and Company, Inco's British subsidiary, this acquisition brings the Company a step closer to being a fully integrated mining and manufacturing operation — extracting and refining ore, producing alloys, and fabricating them into finished products.

Another element in the diversification strategy is a venture capital programme in North America and Europe. Inco will put up money to participate in new and established businesses as in a recent purchase of 10 percent of the equity of the United Tire and Rubber Company of Toronto for a half million dollars. This company produces off-road tires for mining, construction, and forestry vehicles and is the only Canadian-owned manufacturer of tires.

Already the diversification seems to have been justified by the balance sheet. ESB registered record sales of $598 million in 1976. And though 70 percent of Inco's total assets are in primary metal production, 50 percent of its revenues come from other sources. "We intend to continue to grow in the natural resources area, but we probably will grow somewhat faster in other areas," said J. Edwin Carter, shortly before he succeeded Edward Grubb as chairman of the board in April, 1977.[5]

But probably the most important expansion the Company is planning has implications for Canada and the world as yet unclear to anyone.

MINING THE OCEAN FLOOR

The ocean floors are strewn with potato-sized nodules of minerals. At current market prices the estimated value of the nickel and copper in this form likely to be found on the bottom of the Pacific alone is $3 trillion. Inco is one of the leaders in a select group of high-technology corporations researching and developing the means to exploit these rich mineral lodes.

Though these nodules contain 30 percent manganese, their greatest value apparently lies in the nickel and copper which they contain — up to 1.5 percent of each of these materials. According to Donald Sherman, chief resource geologist for the U.S. government, "nickel will be the mainstay of the nodule industry."[6]

The backdrop for this as for most such investments is a mix of technical, economic, and political factors.

Sulphide ore sources are being depleted while the lateritic ores of the tropics are not quite the bargain they seemed before the rise in oil prices. Further, the threat of Third World producer nations forming cartels along the model of the Organization of Petroleum Exporting Countries (OPEC), may worry the corporate and government planners of the industrialized world.* Regardless, alternative sources are always convenient in resisting higher tax and royalty rates for mining.

---

*This threat does not seem real. The attempt to form a copper cartel seems to have failed, and in any case, OPEC has succeeded *with the co-operation and to the benefit of the transnational petroleum companies*.

The problems involved, not surprisingly, are also technical, economic, and political. First, the nodules are located in very deep water and the technology for raising tons of rock from 20,000 feet of sea is in its infancy. Though most of the research has been shrouded in secrecy, it appears that the favoured method consists of sucking the nodules to the surface with something akin to a giant vacuum cleaner. Upon the solution of this and processing problems ultimately depends the economic viability of sea bed mining. But potentially more serious problems — political problems — loom, too.

The United Nations Organization has called the oceans "the common heritage of mankind". But its Law of the Sea Conference has been unable to determine how this heritage is to be governed. Third World countries have expressed the fear that uncontrolled exploitation of the undersea nodules would severely undercut some already fragile national economies, particularly those dependent on nickel and copper revenues. They propose the establishment of an international authority to regulate development and channel some of the resulting wealth to the poorer nations.

The U.S. regards the ocean bed as a source of strategic raw materials which can further reduce its dependence on "unstable" sources in the Third World.* Assistant Secretary of Commerce Tilton H. Dobbin summarized his government's position rather tersely: "We will not allow our national interests and the interests of U.S. companies to be adversely affected by unreasonable delays or a compromise on basic principles."[7]

Unregulated access to resources free of royalties would seem to be the ideal for the mining companies. At last — the perfect investment climate! But to gain official sanction for their activities — and presumably official protection — some ocean mining interests have been pressuring the U.S. Congress to pass laws for the licensing of sea bed ventures.†

Canada is caught in the middle as a country dependent on resource extraction but a relatively rich one, "home" to some of the transnationals involved. At a 1976 session of the Law of the Sea Conference, the Canadian delegation supported a draft proposal sponsored by the Third World bloc which called for the division of growth

---

*Chairman of the American delegation to the Law of the Sea Conference is T. Vincent Learson, former board chairman of IBM. Marne Dubs, ocean research director for Kennecott, has been an "advisor" to the U.S. contingent.
†In 1974, Deepsea Ventures asserted that it had "discovered and *taken possession of*, and (was) engaged in developing and evaluating . . . a deposit of seabed manganese nodules" (emphasis added). It further claimed the right "to take, use, and sell all of the resulting minerals."[8]

in the nickel supply between land and sea mines. Since then, Canada has vacillated on the question of compensation for countries adversely affected by the development of undersea mining and has indicated willingness to accept an initial period of five years in which all growth in nickel demand could be supplied from the sea bed.*

Inco's corporate interest is clearly not identical to Canada's national interest in this matter, at least not in the long run. But the power of the corporations pushing for relatively unrestricted undersea mining is telling. Ocean Management Inc., the consortium in which Inco holds a 25 percent interest, includes Sumitomo of Japan and the AMR group of West Germany. Kennecott Copper has been involved in a project for several years with Mitsubishi of Japan and Rio Tinto Zinc and Consolidated Gold Fields of Britain. Deepsea Ventures operated from 1968 to 1975 under the control of the oil transnational Tenneco, but has been taken over by Ocean Mining Associates, a consortium lead by U.S. Steel. Other corporate giants "actively interested" include Noranda Mines of Canada, Metallgesellschaft of West Germany, France's state-supported CNEXO and Société le Nickel, and the Soviet Union's Marine Geological Prospecting Board.

> *It is more in our best interest to not have a treaty than to have a bad treaty. The spirit of co-operation is lacking and the Third World has very extreme positions and is unwilling to budge in the slightest.... The Third World has a political heritage from British socialism. The countries are anti-American and anti-profit, anti-development, and anti-creation of wealth.*
> — Richard Greenwald, president of Deepsea Ventures

Considering the influence which most of these companies can exert on their host governments, it comes as no surprise that, to quote the *Financial Post*: "Several countries, including Canada and the United States, have warned that they are not prepared to wait indefinitely for an international agreement over the mineral riches, implying the likelihood of unilateral action."[9] The Canadian government, at best, seems eager to offend neither the mining companies nor the mine workers who are concerned about their jobs and the future of their communities. It has been left to the Sudbury City Council and opposition Member of Parliament John Rodriquez (Nickel Belt) to prod the federal government to oppose indiscriminate, unplanned exploitation of the ocean floor.

*Charles Elliot, past president of the Mining Association of Canada and now chairman of Conwest Explorations, is a consultant to the Canadian delegation to the Law of the Sea Conference.

CORPORATE STRUCTURE

Although there has been some diffusion of control over Inco since the 1930s, Inco "is 70 percent controlled outside Canada with over 50 percent of the total ownership located in the U.S."[10] The key interests behind its management continue to be Morgan banks and the law firm of Sullivan and Cromwell.

The original J.P. Morgan and Company was split in two by U.S. legislation which separated commercial and investment banks. The original name was applied to the commercial bank while the investment bank was called Morgan, Stanley Company. In 1959, the former merged with Guaranty Trust to form Morgan Guaranty Trust. Today it is the fourth largest commercial bank in the United States with assets over $25 billion.

As chairman and chief executive officer of Morgan Guaranty Trust, Elmore C. Patterson was a director of Inco for 16 years until his retirement in 1974. The investment bank, Morgan, Stanley and Company, continues to underwrite Inco's stock issues and long-term debt.

Sullivan and Cromwell, one of America's most influential law firms and an early Morgan ally, still continues to be represented on Inco's board, at present by Wm. Ward Foshay. Ashby McC. Sutherland, a senior Company vice-president, is a former member of Sullivan and Cromwell as is Dean Ramstad, a vice-president. Henry Wingate, Edward Grubb's predecessor as chairman of the board and chief officer, had also been with S&C before joining Inco in 1935.

One of the law firm's founders, William Nelson Cromwell, was instrumental in setting up the nickel trust in 1902 and he sat on its board of directors for nearly 50 years. He was among those who advised that Inco "move" to Canada in 1928 to evade American antitrust laws.

The most famous of Sullivan and Cromwell's envoys to Inco was John Foster Dulles. His career exemplifies the function played by corporation lawyers of this type. As Dulles' biographer notes, his early period with Inco was marked by the cultivation of "associations only on the highest financial, political, and social levels."[11] This process climaxed with Dulles' appointment as Secretary of State, a position in which he proved most sympathetic to private corporate needs.

Inco is linked to several other of America's dominant corporations. It shares with the Rockefeller's Exxon Corp., the largest petroleum company in the world, a longstanding relationship with Sullivan and Cromwell. Until recently J.K. Jamieson, chairman of Exxon, sat on Inco's board and in the past, Laurence Rockefeller, grandson of the founder of the Standard Oil empire (Exxon's predecessor) was an Inco director. C.F. Baird, a former Exxon executive, is the new president of Inco and has been vice-chairman of the Company's board in charge of finances.* Robert de Gavre, now Inco's Treasurer and the man responsible for organizing financing for the company's overseas expansion, is another Exxon old boy.

While the Rockefeller connection facilitates relations with the Chase Manhattan Bank, the third largest in the United States, another board member links Inco with the Bank of New York, a moderately sized bank with assets of $4.2 billion. Samuel H. Woolley, who has been on the Inco board for many years, continues to be listed as the "former chairman" of the Bank of New York.

Another key connection which Inco has made recently is with the New York based Citibank. Formerly the First National City Bank, this powerful financial institution was the prime mover in financing Inco's operations in Indonesia and Guatemala.

Inco's principal connection in Canada has been with the Bank of Montreal and the group of companies associated with it.† When Inco was being put together around the turn of the century, the Bank of Montreal was the most important financial institution in Canada as well as having close links with the Morgan interests. Its involvement in the nickel trust was quite logical. In recent years it has been outstripped by two other Canadian banks, the Royal and the Commerce, in the accumulation of assets, but the Montreal remains extremely influential because of the corporate connections which it has maintained. This network of connections evolved from the political and economic strategy which lead to Canadian Confederation. Put simply, the Bank of Montreal was one of the more important interests behind the building of the Canadian Pacific Railway and the formation of a unified Canadian state.

The link between Inco and the Bank has been symbolized by interlocking directorships — the chairman of each company has held traditionally a seat on the board of the other. This remains true even today with Arnold Hart, chairman of the executive committee of the Bank, and Edward Grubb of Inco. Now that the latter has retired there

---

*Baird is a former Under Secretary of the U.S. Navy. His current position is perhaps appropriate considering the importance of nickel for military purposes and considering the importance U.S. Navy contracts had in boosting Inco's predecessor, the Canadian Copper Company.

†The Molson Companies (Toronto), Rolland Paper (Montreal), the Steel Company of Canada (Toronto), and James Richardson & Sons (Winnipeg) all have interlocks with the Bank of Montreal and representatives on the board of Inco. Canadian Pacific Railways and Bell Canada, two firms generally associated with the B of M and Morgan Trust Company of Montreal, have had representatives on the Inco board in the recent past.

is every reason to expect that his successor, J. Edwin Carter, will replace him on the board of the Bank.

The interlocks between Inco and Canadian finance extend to two other chartered banks. During the Sixties, while the Company's chairman sat on the board of the Bank of Montreal, its president served on the board of the Toronto-Dominion and its executive vice-president on that of the Bank of Nova Scotia. The significance of such connections is indicated by the loan arrangements for P.T. International Nickel Indonesia. Each of these three banks has underwritten part of the capital debt with the Bank of Montreal leading a syndicate loan.

In his study of Canada's dominant class, Wallace Clement calculated that the average resource corporation has 13.3 directors.[12] Inco has 22 members on its board, a recent reduction from 25. This suggests a policy of maintaining a broad power base within the Canadian corporate elite. Two particularly influential directors are George Richardson and Peter D. Curry. Richardson, brother of former Minister of Defence and Liberal Member of Parliament James Richardson, heads the Winnipeg-based family company built on the grain trade and finances, and is governor of the Hudson's Bay Company. He holds 31,000 shares, more than any other Inco insider. Curry is president of the Power Corporation, the conglomerate with interests in shipping, insurance and finance, communications, real estate, and manufacturing which its chairman Paul Desmarais has parlayed into the most powerful corporate complex in Canada.

While important financial and marketing functions are still conducted in New York, Inco has moved most of its top executives to its offices on the 44th floor of the Toronto-Dominion Tower in Toronto. As Edward Grubb has observed: "Things can be done better on the scene than by remote control."[13] This move may have made easier borrowing from Canadian financial institutions and it has given the Company's top managers greater access to federal and provincial government officials.*

One forum which Inco has used to make its already "resonant" voice even more clearly heard, is the Conference Board of Canada. Grubb has served as a director of the Canadian Council of the Conference Board, a branch of the American organization of the same name which is an elite think tank intended to influence government through research and policy advice. The importance and prestige of

---

*While some may quibble over whether more Inco shares are held in the U.S. or in Canada, the essential fact is that the most powerful interests are privately held in large blocks. Over half of Inco shares are held in blocks of 100,000 or more.

the Conference Board is indicated by the appointment in May, 1976 of A.J.R. Smith to an Inco vice-presidency in charge of public affairs. At the time of his appointment Smith was president of the Board and previously he had been chairman of the government-appointed Economic Council of Canada.

The C.D. Howe Research Institute, formerly the Private Planning Association of Canada, is another body similar to the Conference Board to which both Grubb and Smith have connections. It tends to organize more factual research on economic problems and less surveying of management opinion than the Conference Board and so the two function well together. Smith is a former secretary-treasurer of the Institute while Grubb is a member of its Canadian Economic Policy Committee.

The changes which have occurred in Inco's management in the past decade are significant for the Company but of little concern to those outside its direct influence. More Canadian capitalists have been added to the board of directors while the British influence has been diminished. (This erosion has been gradual since the merger with the Mond Company in 1928 and has reached the point now where only one director, the Rt. Hon. Lord Nelson of Stafford, is based in Britain.) Though they may live in the elite enclaves of Toronto, top Company managers continue to be American. Edward Grubb was from New Jersey and J. Edwin Carter is from Georgia (though he is not related to the other Carters of Plains, Georgia).

To those people dependent on Inco for jobs particular management decisions can be important. Efforts to increase production and reduce costs may endanger the jobs of some and the health and safety of others working in the mines, refineries, and smelters. Efforts to diversify sources of supply around the world and under the sea may weaken the resolve if not the power of governments to attempt to claim for the common good part of the wealth created in nickel production. For in the final analysis, the ability of the Company to choose its sources from among various countries in the short run, gives it the power to "go on strike" against any government which "gets out of line".

But none of these decisions depend on the goodwill or lack thereof, of the people who make and execute them, or on whether Inco is a good or bad "corporate citizen". Though there may always be choices to be made, the bounds within which they are located are dictated by the larger political economic system. Choosing wisely within those bounds is the test of a good corporate management; but choosing within them is the only option of management in a private corporate system.

Inco: Selected Subsidiaries

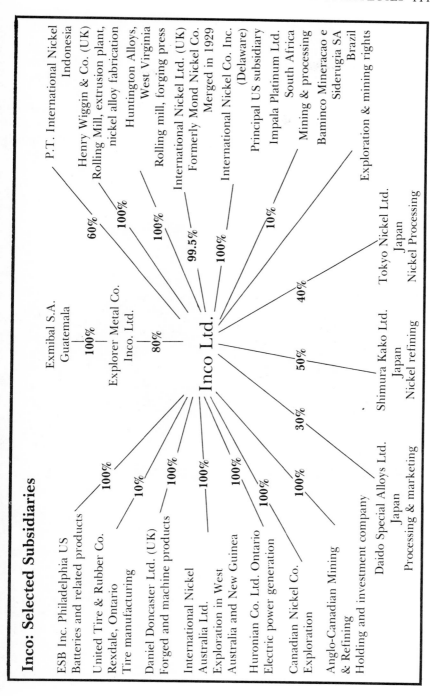

Scorched Earth...

Anyone who has travelled the Trans-Canada Highway through Northern Ontario has passed through the Sudbury Basin, the prime source of Inco's immense wealth. Most Canadians associate this area with blackened rock outcrops and barren landscapes — a treeless vista which has earned the district fame as the testing ground for moon-travellers. But Sudbury has not always been so foreboding. Its surrounding terrain once sustained thriving lumbering and agricultural industries.

It is difficult to place a dollar value on the environmental damage caused by Inco over more than seven decades. But an unpublished report, prepared for Environment Canada in 1974, provides a staggering cost estimate of the damage from sulphur dioxide (SO_2) to the Sudbury Basin. This confidential document, entitled *The Sudbury Pollution Problem: Socio-Economic Background*, reckons damage by a complex but cautious formula originally developed by the United States Department of Health, Education and Welfare. It puts the price for environmental damage at "approximately $465,850,000 caused by emissions to human health, vegetation and property value in the Sudbury area on an *annual* basis."[1] (Emphasis in the original). The total included Falconbridge emissions, but as Inco accounted for 84 percent of the total, its share of the damage costs was roughly $391,314,000 — well over the Company's profits for any year in its history.

The Report's authors are careful to emphasize the conservative nature of their estimate — it does not include the effects of SO_2 on animals, soil, and the aesthetic appeal of the natural environment. They merely applied a formula which calculated that each pound of sulphur dioxide emitted in the U.S. in 1968 cost that society 12.5 cents. Adjusting to take account of the lower population density of Sudbury and five years of inflation, the cost was calculated at 12.1 cents per pound in 1973.[2]

114 THE BIG NICKEL

Low level emissions from inside the various plants were not included. But an estimate was made for the damage caused by particulate emissions from the Inco superstack — that is, the dispersal of particulates as distinct from gas — of $4,311,853.[3]

These figures can convey nothing of the human and environmental debilitation caused by Inco's pollution. They do shed a different light on the Company's public relations platitudes concerning its huge tax bill, its massive payroll, its immense expenditures on Canadian goods and services, and the benefits these all bring to the country through the "multiplier effect".*

Sulphur dioxide is one of the most thoroughly studied and best known air pollutants, probably because it results from the burning of the many fuels which contain sulphur. It is also the most important pollutant released in the refining of the sulphide ores which contain nickel.

A pungent, colourless gas, SO_2, even in small concentrations appears to contribute to lung diseases such as bronchitis. High concentrations are fatal and the gas is suspected of being carcinogenic — that is, cancer-causing — in humans and other animals. When sufficient sulphur dioxide is absorbed into plant tissue, the organism usually dies. If the plant is not killed directly, it is often rendered unusually susceptible to damage from bacteria, fungus, and insects. White pine, a common species in Northern Ontario, is especially prone to SO_2 poisoning.

Airborne sulphur dioxide mixes with atmospheric moisture to produce what is known as "acid rain". Such precipitation entering soil, diminishes fertility by leaching out valuable nutrients. In lakes, rivers, and streams, it adversely affects aquatic life.

The failure of the Canadian government to release this document is not surprising. Federal and Provincial governments through their inaction have allowed this damage to proceed unabated for decades. Indeed, the report itself concludes that residents of Sudbury have "had the general feeling that there is collusion taking place between Inco and Falconbridge and government officials."[4]

THE EARLY YEARS
In 1886 the Canadian Copper Company, Inco's forerunner, started open bed or heap roasting, as a primitive method of smelting. This process simply involved dumping raw ore onto layers of cordwood,

---

*This "effect" is an attempt to compute the ultimate result of an investment creating jobs which in turn create demand for goods and services, and so on, throughout a year. Mining companies seem particularly prone to not only taking credit for what is beyond their control, but also exaggerating this effect.

setting the mass aflame and letting it smolder for several months. Gradually the sulphides were separated from the ore as an acrid sulphurous gas. Since no machinery was required and wood was still plentiful at that time, this was a cheap method of reduction. But it was also extremely inefficient, for it allowed some of the nickel as well as copper, gold, silver, and platinum by-products to be washed into the soil by the rains.

Some apologists say (cutting down sulphur emissions) can't be done. Nor it won't be done, as long as we fellows grin and bear it — as long as we keep on spitting and coughing. These companies should be indicted tomorrow for maintaining a public nuisance.
— a Sudbury area farmer, 1916[5]

The effect of open bed roasting on the environment was nothing short of catastrophic. Although the Sudbury Basin had undergone extensive "bare-ground" logging for railway ties, mine pit props, and even for lumber exports to the U.S., the toxic sulphurous fumes added to the destruction of the remaining forest and precluded the possibility of natural regeneration.

Even the smallest blade of grass could not survive, as an early traveller to the region observed: "A more desolate scene can hardly be imagined than the fine white clay or silt of the flats, through which protrude, at intervals, rough rocky hills, with no trees or even a blade of grass to break the monotony."[6]

The sulphur dioxide killed not only the natural vegetation of the area but wrought an equally disastrous effect on agriculture. Farmers located in townships near the Copper Cliff roasting beds watched helplessly as their crops were ruined year after year. In 1906, one angry resident asked in a letter to the Sudbury *Journal* if the citizens were, "... for the benefit of the International Nickel Company, to be deprived of the benefits of fresh nature?"[7]

By 1916 the region's farmers had had enough. Six hundred thousand tons of SO_2 were being emitted annually by the nickel companies and the farmers had grown tired of plowing their crops into the ground. They sent a deputation to the district board of trade and petitioned the courts for damages.

Many of the farmers had bought their land from the government of Ontario with the stipulation that it be brought under cultivation. But by 1916, the Province had decided to withdraw 12 townships from settlement on the grounds that they were "unfit for cultivation". It was obvious that the government, rather than acting against Inco for destroying the productive capacity of the farmland, had chosen to accommodate the corporate interest.

J.H. Clary, a farmer from Dill township, claimed that the farmers even lacked sufficient grain to sow new crops. He labelled the claim that the land was unfit, as "a damnable lie.... Why were those townships withdrawn? Simply to allow the two big smelting companies to use the most primitive methods of treating ore, and also the cheapest, so as to increase their profits."[8]

After a spate of lawsuits against Inco by angry farmers, the Legislature of Ontario passed in 1921 the Damages by Fumes Arbitration Act. This Act established a government-appointed arbitrator who was empowered to make awards for damage to crops, trees, and other vegetation by sulphur dioxide. However, the judgement of the arbitrator precluded *all* other legal claims which victims of such damage could make normally under law.[9]

In 1924, the Damages by Fumes Arbitration Act was amended to give the Minister of Mines power to overrule the arbitrator, with no further right to appeal. This law, which effectively gave Inco legal sanction to defoliate and in other ways destroy vegetation on land which it did not own, was finally repealed in 1970.*

A second governmental "solution" to the pollution problem was provided by "smoke easements" under Ontario's Industrial Mining Lands Compensation Act, which is still in effect. Smoke easements were secured by payments from Inco or Falconbridge to owners or lessees for damage from sulphur dioxide to their property. The easements bound all future owners and lessees of the property covered to the original easement. Land titles in the Sudbury Land Registry Office still contain records of these agreements, even though titles may have changed hands ten times since the original cash settlements were made. The Environment Canada Report recommends a study of the legality of the entire smoke easement system "and the denial of common law remedies" they entail.[10]

Although the smoke easements have long exempted Inco from damage suits from land owners, the Arbitration Act did force it to make token payments for its air pollution. Since the end of World War II Inco has been paying around $50,000 annually to farmers in the Sudbury district. In 1974, fifty-eight years after the first major protest, the Company was finally convicted for polluting the air, on a charge brought not by the Crown but by the Sudbury

---

*Also repealed in 1970 was a 1942 Ontario Order-in-Council stipulating that individuals granted Crown land could not sue mining companies for damage to the land by sulphur pollution.

Environmental Law Association. The penalty? A fine of $1500.

THE SUPERSTACK: DIFFUSION AS A SOLUTION

In 1967 Inco's annual report included for the first time a section entitled "Pollution Control". In a masterful understatement, the Company allowed that: "Pollution of the environment is not a new problem...." But in the late Sixties it was a "new issue" of public concern and both government and industry were feeling pressured to make the right gestures.

Acting under the amended Ontario Air Pollution Control Act of 1967, that province's Minister of the Environment, George Kerr, issued a Control Order to Inco on July 13, 1970. The Company was instructed to gradually reduce its SO_2 emissions from 5200 tons to 750 tons per day, by December 31, 1978. This reduction was to take place at a pace sufficient to have brought the level to 4400 tons daily by the end of 1974 and 3600 tons by year's end 1976.

The same decree required a reduction in the emissions from the Company's Iron Ore Recovery Plant (IORP) to a limit of 250 tons per day by December 31, 1972. Finally, the Order demanded the construction of a 1250 foot stack to replace the existing chimneys at Copper Cliff.

To the public, it appeared that the Ontario government was at last cracking down on one of the Province's worst polluters. But Eli Martel, member of the Provincial Parliament for Sudbury East, maintains that the Company built the "superstack" primarily because it had to replace its smelter's two rapidly deteriorating chimneys. A check of Inco's annual reports does reveal that plans for the new stack were first announced in February, 1969, a full seventeen months before the Ministerial "Order".[11]

In 1972, the big stack was completed at a cost of $13 million. Heralded as the tallest chimney in the world, it successfully diffuses Inco's sulphur dioxide well beyond the Company's immediate operations. But even the deadline for its completion had to be extended by the government from the end of 1971 to August 31, 1972.[12] The Company also managed to get an extension of fifteen months for one of the emission deadlines.[13]

In granting this second extension, then Environment Minister, James Auld, noted that "Ontario operations are feeling economic pressures from off-shore nickel deposits".[14] He did not bother to mention that even in the recession year 1971 Inco's profits had topped $94 million and that the year before they had been $209 million. Nor did he note that Inco itself was developing those off-shore deposits which he feared might pressure the "Ontario operations" of the Company.

Floyd Laughren, the New Democratic MPP for Nickel Belt, was moved to comment on this relaxation of the Province's regulation: "I would assume that pollution control is necessary due to the ill effect of pollution — not because it is economical to industry.... Either the emission of 650 tons of SO_2 per day is harmful to health or it is not."[15]

> *Government has been extremely lenient with Inco and Falconbridge. Historically there have been no prosecutions under applicable environmental legislation, and from 1924 to 1970, there was a curtailment of a citizen's right to sue for pollution damages, and there has been a lack of government-sponsored research on the damage caused by the copper-nickel smelters.*
> — The Sudbury Pollution Problem (Environment Canada, 1974)

Government leniency has extended not only to the scheduling but also to the financing of the Company's pollution clean-up. Under the federal government's Accelerated Capital Cost Allowance Programme, Inco was granted a series of tax deductions between 1970 and 1974. The total write-off amounts to $7,928,773, a tidy sum for one of Canada's more profitable corporations.[16] The Ontario government has also saved Inco $398,972 by exempting pollution abatement equipment from its sales tax.[17]

WHAT GOES UP MUST COME DOWN

The effects of the superstack on the local Sudbury environment have been indisputably positive. Sudbury residents, long the victims of Canada's most severe SO_2 pollution, were proud of a 1975 survey which determined that their city actually had the cleanest air of any in the country.[18] The effect on vegetation has been equally dramatic. In some places, sparse growth is returning for the first time in decades.

But it would be a mistake to assume that the "greening" of Sudbury will continue apace. Keith Winterhalter, a biologist at Laurentian University, has made a careful study of re-vegetation in the district. He predicts that it could be generations before plant life returns to normal because years of acidic and metallic deposits have accumulated in the region's soil, rendering it infertile for even the simplest plant growth.[19]

Erosion has been a second critical problem — in many areas the soil has been washed away due to the lack of plant life. These problems are not insurmountable but their solution requires large and costly applications of lime, fertilizer, and topsoil. Using this mix, Winterhalter has had great success with small experimental plots of hardy grasses even in the bleakest locations.[20]

While the effects of the stack locally may be positive, the broader effects are unknown. By dispersing the effluents high into the prevailing winds, the sulphur dioxide is made to seem to disappear. It is, though, mixed in with the industrial wastes of east central North America, so that it becomes difficult to differentiate Sudbury's waste from that of Detroit, Pittsburgh, or Cleveland.[21] Studies by the Ontario Ministry of the Environment have tentatively concluded that the direct effect of Inco's superstack emissions within a hundred mile radius of Sudbury are minimal.[22] But environmentalists are becoming increasingly concerned about another more insidious result of waste SO_2 — acid rain.

Around 98 percent of the sulphur dioxide produced in Sudbury is now dispersed from the immediate area. But that still worries the head of the botany department of the University of Toronto, Dr. T.C. Hutchinson. "We can't seem to win, actually. We start to clean up and someone else gets it," he notes.
      — quote from the Toronto *Globe & Mail*, Feb. 28, 1977

The product of sulphur dioxide mixing with moisture in the upper atmosphere, acid rain contaminates both the soil upon which it falls and the watershed of that soil. As water becomes more acidic, it can no longer support the small organisms that provide the basic nutrition for aquatic life. Lakes in New England and Canada are now beginning to show the destructive effects of acidification.

Professor Ross H. Hall, Chairman of the Department of Biochemistry at Hamilton's McMaster University, noted in an article on the superstack the long-term dangers of sulphur in the atmosphere: "Recent studies have shown that man is now contributing about one-half as much as nature to the total amount of atmospheric sulphur compounds and that by the year 2000, at the present rate of expansion of industrial plants, he will be contributing just as much as nature. In North America he will be contributing much more than nature. Nature's cycles are delicate and designed to accommodate only so much sulphur."[23]

Obviously sulphur must be extracted during the production process. The technology for its recovery has been available for 40 years and is currently in use, notably in the sour gas fields of Alberta. While sulphur has commercial value as fertilizer, over-production in recent years has driven its price down, rendering its recovery and sale unprofitable for companies like Inco.

The emission target of 750 tons per day set by the Ontario government, to be achieved by Inco by the end of 1978, is a considerable improvement over the pre-stack level of 5200 tons per day. But in the spring of 1977, the Company and the Ministry were still discussing the 1978 target behind closed doors, arousing suspicion that Ontario's Progressive Conservative government might grant yet another extension.[24]

LINGERING LEGACY

It is difficult to put a price tag on the damage that has been done to the image of Sudbury in Canada, but there is little doubt that ecological destruction continues to cost the city dearly. Two of the three mainstays of the Northern Ontario hinterland economy — forestry and tourism — have been denied the Sudbury Basin. The productive forests which exist elsewhere in the north are just no longer found here. Likewise, sulphur-blackened granite is not attractive to tourists. Fishing camps which once thrived have suffered drastic losses in clientele as fish stocks have declined. While water quality is improving and lakes are being re-stocked, it may be years before fishing and hunting can support a viable industry.

In the late Sixties, 150 lakes in the Sudbury airshed were tested by federal authorities. Thirty-three were dead, totally void of aquatic life, and another 37 were deemed "marginal". An Ontario government report in the same period estimated that at least 25,000 acres of pickerel fishing water and 17,000 of lake trout water had been completely lost.[25] These losses have been most directly felt by tourist camp operators. The Peloquin family watched their once successful business on Lake Chiniguchi, northeast of Sudbury, steadily decline. In 1968, an American angler who had been a regular visitor to the Peloquin camp, wrote to the Ontario government to complain about the deteriorating situation. "He was told that nothing would be done about Lake Chiniguchi until it became profitable for the mining companies to clean it up."[26]

In 1974, the Sudbury Regional Development Corporation was established. Despite an expensive promotional film and glossy publicity brochures, the Corporation has yet to attract even a single major secondary industry to the area. One of the difficulties would seem to be the deep-seated aversion many Canadians still feel towards the city's physical setting. Even with its immense natural wealth, its location on major transcontinental rail and highway routes, and its skilled and stable industrial workforce, Sudbury remains a one-industry town. At work and at home, the residents of the Sudbury area are still paying the price for decades of Inco's irresponsibility.

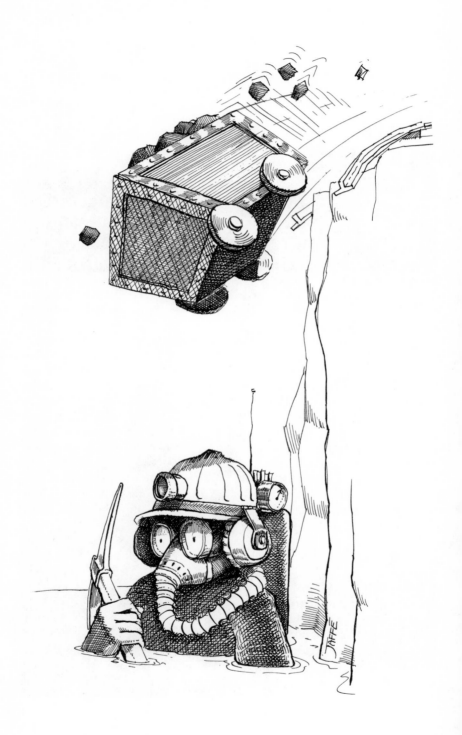

... And Broken Bodies

In its 1945 pamphlet, the Romance of Nickel, Inco glorified "the prospector and the engineer, the scientist, the statesman, the salesman, the investor and the business executive". The Company owed its status to these who had "contributed their life's work, their skill, their wealth and their knowledge".[1]

To whom did the mine and smelter workers of Sudbury owe their lot in life? From the earliest days of the operation of the Canadian Copper Company, Inco's forerunner, the pages of the Sudbury newspapers have recorded death in the nickel mines: Dec. 1, 1891 — Mining fatality at the Canadian Copper Co.; Sept. 8, 1892 — Five killed at Blezard Mine by rock burst; Sept. 26, 1901 — Accidental death in roast yard of Canadian Copper Co.; March 13, 1902 — Trammer killed at Frood Mine; July 24, 1909 — Three killed at Creighton Mine site; Sept. 23, 1909 — Worker killed by train in Copper Cliff smelter; Feb. 17, 1910 — Worker crushed by falling rock at Creighton Mine. (All examples are from the Sudbury *Journal* of the date given.)

And on and on. But neither the *Journal* nor its successor, the *Star*, were able to recount the daily reality of work life for the thousands who did the bidding of the engineer and the business executive and who actually grappled with nature to produce such wealth.

A RISKY BUSINESS

By its inherently destructive nature, mining involves exceptional hazards. Blasting thousands of feet into the earth and tunnelling for miles under it, inevitably lead to encounters with the intense pressure of the constantly shifting crust of the planet. "More than 1,000 miners have died in on-the-job accidents in this decade at Canadian mines."[2] This incidence of fatalities makes mining the most dangerous job in the country.

It is not just what mining intrinsically entails, but also how it is done that makes it risky. The technology used in hardrock mining tends to be designed for maximum production efficiency rather than maximum safety. Blasting operations produce gaseous compounds and fine dusts. Drilling adds more substances that might be inhaled. Scooptrams, bulldozers, trains, and drills emit gases, heated oil mists, and tiny particles, as well as making what can be a literally deafening noise.

> *It is a mathematical and practical impossibility to simultaneously maximize two variables such as occupational health and safety and industrial production and profits. Given the current state of affairs, an increase in the level of one usually results in the decrease of the level of the other.*
> — Robert Sass, "The Underdevelopment of Occupational Health and Safety in Canada: Contradictions and Conventional Wisdom"

These conditions, in themselves a problem, combine with the intense heat deep below the earth's surface and the poor lighting common to underground operations to create strain and fatigue in the miners, which increases the likelihood of accidents.

Above ground processes to which nickel ores are subjected pose their own peculiar threats to health, while at the same time giving cause to many of the safety problems common to large scale, highly mechanized industry. Concentrating processes produce metal dusts as well as such toxic compounds as sulphur dioxide, nickel carbonyl, and carbon monoxide.

In a 1976 publication called *Health and Safety of Nickel Workers*, the International Metalworkers' Federation came to this general conclusion: "It is almost impossible to overstate the health hazards associated with the production of nickel. First, nickel itself has been shown, in a wide range of tests in animals, and in human epidemiology to cause cancer. In addition to cancer of various sites of the body, nickel production has been shown to cause silicosis, lung fibrosis, heart disease, skin disease, eye damage and a vast array of other maladies."

Historically, the mining industry — Inco is Canada's largest mining company, employing nearly half Ontario's mine workforce — seems to have ranked health and safety far below production on its list of priorities. Little or no compensation was available in the early days of mining to injured workers and industrial diseases went totally unrecognized. The sick and disabled faced unemployment and debt from medical expenses. The companies usually found it easy to replace the victims of their operations.

However, as the pressure for unionization mounted, both industry and government had to concede some attention to these problems.

The Workmen's Compensation Board (WCB) was established by the Province in 1916 and the companies had to contribute financially to the low level insurance against injury thus provided. In 1930, the Mines Accident Prevention Association of Ontario (MAPAO) was created under the umbrella of the WCB.

The MAPAO illustrates the extent to which industry and government have been jointly responsible for the health and safety conditions in the mines of Ontario. Though it is funded by the WCB, it is staffed at the upper echelons by industry representatives. The executive director of MAPAO is Norman Wadge, previously a public relations officer for Inco, while his administrative manager, John Ridout, is another former Inco employee who had specialized in industrial relations.* Though there has been no worker or union representation in MAPAO, the provincial government has been content to leave most questions of health and safety to this organization. Most unionists feel that the corporations have been left to regulate themselves. In the words of one union official: "This is akin to putting Dracula in charge of the blood bank."[3]

The Mines Engineering Branch of the Ministry of Natural Resources, the government office responsible for ensuring the safe operation of mines, has repeatedly come under attack for alleged collusion with the mining companies. This branch, like MAPAO, is staffed mainly by people who have worked for the mining corporations. It would seem difficult for such functionaries to act as neutral observers in situations which involve their former employers. This general tendency of Ministry staff to show leniency toward the companies is sanctioned by the Ontario Mines Act itself — not one regulation in it deals with health and safety.†

Ever since unions in the mining jurisdiction achieved some stability in the Forties, they have combined efforts to improve working conditions through their collective agreements and attempts to pressure government to regulate health and safety standards industry-wide. In recent years the amassing of compelling evidence on the extent of industrial disease has given added weight to the demands by the United Steelworkers and other unions for government action.

The relationship between certain occupations and the incidence of cancer has been particularly dramatic in drawing public attention to official complacency.

*Wadge was succeeded in the summer of 1977 by James M. Hughes, an engineer who worked for Inco prior to joining the Ontario Ministry of Natural Resources.
†The Employees Health and Safety Act, (Ontario), passed in 1976 following the Report of the Ham Commission, supercedes the Mining Act.

> *There is increasing evidence that most inorganic nickel compounds cause nasal (nose) and lung cancer in exposed workers. Approximately 250,000 workers use nickel metal and nickel salts in industries that produce nickel alloys and stainless steel or do electroplating.*
>
> *Information on the rate of nasal and lung cancers comes from international sources.*
>
> | | Number of expected cancers | Number of observed cancers |
> |---|---|---|
> | Inco refinery (Wales) | 2.3 nasal | 56 |
> | | 27.4 lung | 145 |
> | (Port Colborne, Ontario) | 0.47 nasal | 24 |
> | | 40.9 lung | 76 |
>
> *These two studies show a significant difference in the number of expected cancer cases and those actually seen in the workers.*
> — Health Hazard Alerts, *Monitor*, May-June, 1977
> Centre for Labor Research and Education, Berkeley, Calif.

One notorious example is Inco's sintering plant in Sudbury. The sintering process in use throughout the 1950's converted nickel sulphide to nickel oxide at high temperatures. It has since been recognized as the key factor contributing to the high incidence of respiratory cancer among those who worked there. Another example is the high level of cancer among those exposed to ionizing radiation in the uranium mines at Elliot Lake.

In 1974, the long complacent government of Ontario established a Royal Commission on the Health and Safety of Workers in Mines. The lone commissioner was Dr. James Ham, dean of graduate studies at the University of Toronto. This Commission was a valuable forum for unions and individual workers to bring forth and document the grievances which they had harboured relating to both accidents and industrial diseases. In large part, it was an opportunity seized to record both general conditions and specific problems.

One Inco worker testified that the temperature in Creighton number nine, the deepest shaft in the Western hemisphere, reached as high as 130 degrees Fahrenheit. "We sweated while we ate our lunch," he noted, casually adding that one government inspector who did not find the environment so convivial, had to be carried out of the mine.[4]

Gib Gilchrist, a USWA official, pointed out to Dr. Ham that the union at Inco in Sudbury, local 6500, has made an average of 1200 safety complaints annually to the government, resulting in only two attempts by the Ministry to prosecute the Company.[5]

The Ham Commission confirmed many of the charges made about the behaviour of both companies and government on health and safety concerns. It branded as "paternalistic" and "unacceptable" the failure of managers, civil servants, and politicians to inform workers properly of some of the daily dangers, particularly those not so obvious, that must be faced in mine complexes.[6] Further, Dr. Ham observed that under the present system "the worker as an individual and workers collectively in unions, have been denied effective participation in tackling (health and safety) problems".[7]

Recognizing the inadequacy of a system in which corporations are expected to regulate themselves, the Ham Commission advocated worker participation in policing health and safety standards. On a more general level, it compiled detailed evidence of the human price paid for neglect and inaction.

INDUSTRIAL ACCIDENTS

In the period 1970-1974, Inco had the dubious distinction of having one of the worst safety records of all Ontario mining corporations.*[8] In the three years years since, the incidence of serious, non-fatal accidents in the entire mining sector has increased at a disturbing rate largely as the result of increased accidents in nickel mining, particularly at Inco.[9]

Between 1960 and 1974, seventy-one workers were killed at International Nickel's Sudbury operations.[10] Prior to 1972, coroners' inquests into fatalities at Inco were rather tidy affairs, usually involving the local coroner, one official from the provincial government, a member of the Company's private police force, and the superintendent of the department in which the death had occurred. Worker participation was discouraged and union representatives had difficulty contributing to the proceedings as they were often barred from the scene of the accident. As if symbolically, inquests were frequently held in the office of the superintendent involved.[11]

Pressure from the union eventually forced revisions to the Coroner's Act which require more thorough investigations. A special Inquest Committee set up by USWA local 6500 is now able to ensure at least minimum standards of impartiality. Nevertheless, in 1976 four workers died at Inco's Frood Mine, one of the Company's 18 mines in Canada.

Not all mine accidents are fatal, obviously. They range from minor scrapes to mishaps involving haulage equipment, to injury from falling rocks and timbers, and electrocution. While it is often impossible to

---

*Only Chromasco Corp. and the tiny Ross gold mine had worse records in this period.

isolate their precise causes, they usually arise from a complex interaction of factors, including both poor working conditions and unsafe acts. The period in the early Seventies with high accident rates at International Nickel coincides with that in which several thousand workers were dismissed from the Company's Ontario payroll. After a poor profit year, the Inco Board appointed Edward Grubb president in 1972, giving him a clear mandate to cut costs. Dubbed by *Fortune* Magazine a "no-nonsense broom-wielder", Grubb accomplished his task by eliminating thousands of jobs while maintaining output.

Total Non-fatal injury experience by metal group and selected corporation in Ontario

	Average frequency of non-fatal injuries per million man-hours 1970-74
Gold	28.24
Dome Mines Ltd.	39.32
Kerr-Addison	17.03
Iron	22.74
Algoma Ore	20.83
Sherman Mine	31.08
Copper	14.36
Noranda Mines — Geco Division	19.97
Texasgulf Canada Ltd.	5.34
Silver	55.12
Agnico Eagle Mines Ltd.	55.88
Nickel	54.76
Falconbridge	21.53
Inco	62.88
Uranium	20.68
Rio-Algom Mines Ltd.	21.53
Miscellaneous Industrials	33.24
Canada Talc	39.39
Sobin Chemicals	35.76

Source: Report of the Royal Commission on the Health and Safety of Workers in Mines, Table D7.

Switchmen who used to precede ore trains to warn of danger, were removed — trains are now run blind from the rear.[12] Trackmen who maintained the underground tracks and level bosses who kept up the physical plant at the various mine levels were also dismissed.[13] Experienced miners cite these cutbacks as the main cause of increased accidents.

The Ontario Royal Commission pointed to inadequate training of miners as one basic cause of injuries. It determined that workers new to the job had a greater chance of being injured or killed than their more experienced colleagues.[14] Because there are no laws regulating the training of miners and because of the high turnover in the labour force, many of those working in mines are untrained and inexperienced.*

> *A lot of reported accidents are not accidents at all. The men fake them because they don't want to work.*
> — D. Anderson, superintendent of Inco's Levack mine. This mine had the highest average frequency of accidents in Ontario mines during the 1970-1974 period.

The cause of high turnover itself remains the subject of dispute. The more myopic or perhaps misanthropic commentators attribute it to alleged abuses of unemployment insurance. The real causes seem more likely to involve the relative isolation of most mining communities and the dangerous nature of the work. The turnover could just as well be a result rather than a cause of accidents.

One failure of the Ham Commission was its ignoring the bonus or "incentive" system as a cause of accidents. Most mining companies, Inco included, offer bonuses in a system basically comparable to "piece-work". The extra $500 a month that can be made "on bonus" not only attracts workers underground in the first place, it keeps them working there at a rapid pace, unsupervised. Working fast often requires taking short-cuts and ignoring safety measures. As one Steelworker official put it: "Bonus is the company's way to make the miners break the rules."[15] In the event of an accident, the blame can easily be attributed to "employee carelessness". The companies can emphasize production above all else without worrying about the means — safe or unsafe — of attaining this goal.

Though mining unions have been reluctant to oppose the incentive system because of the potential for disputes within their memberships over the matter, they are increasingly exposing the consequences of it. An example from the *Miners' Voice*, a USWA publication, is a report on fatal accidents at the Inco mines in Thompson, Manitoba: "June was another month, and miners at Thompson were getting edgy. Every month this year the big Inco Ltd. nickel operation had recorded one death.

*Manitoba has a certification programme for miners comparable to other trade apprenticeships. It is not compulsory that all miners undergo it and until it becomes more widely established and the incentive to take it increases (i.e. wages are raised), it risks becoming marginal.

"In May miner Zoran Nikolic was killed when an 11-foot-long section of rock over his head fell and crushed him.

"In April Guy Elgar was killed in a similar accident in the same No. 3 mine....

"... Like Nikolic, the 24-year-old Elgar was a miner with some experience — but both were on the bonus system....

"Companies claim they need the incentive payments based on production to get miners to work underground, where supervisors can't visit some areas more than once a shift. So the bonus system becomes, as one union official says, 'the voice of the boss in the dark of the mine'."[16]

INDUSTRIAL DISEASES

Research into industrial diseases is sadly underdeveloped. Although there are approximately 15,000 toxic materials used or produced by industry today, only 500 have a scientifically established Threshold Limit Value — the level at which exposure becomes dangerous.[17] Today in Canada, it is still not possible for a physician to become a certified specialist in occupational medicine. As usual, the medical establishment is emphasizing curative as opposed to preventive medicine in dealing with job-related illnesses.

The human lung was not made to function in industrial society.
— R. Sass, "The Underdevelopment of Health and Safety in Canada: Contradictions and Conventional Wisdom"

Workers in integrated mining complexes are exposed to a variety of dangers — some common to most heavy industry, some peculiar to the mining industry in general, and some confined to nickel mining in particular.

Nickel ores, especially Canada's sulphides, tend to be low in actual nickel content and are often very complex. In the numerous reduction processes necessary to produce commercially useful nickel, a wide range of elements and compounds are released, many of which are extremely toxic. Those working in smelters and refineries run the risk of contracting a variety of serious illnesses including cancer. What follows is a brief list of these toxic substances: 1) nickel compounds themselves. Nickel has been well-researched and in many nations the element and its compounds are recognized carcinogens. Nickel carbonyl, a process gas, is extremely toxic; 2) metals found in nickel ores. These include cadmium, arsenic and chromium, all carcinogens; cobalt, a recognized animal carcinogen; and selenium and lead, dangerous toxins; 3) substances used or released in nickel processing.

Carbon monoxide, hydrogen sulphide and potassium xanthatate are highly toxic, hydrogen fluoride is highly corrosive, and sodium sulphide is irritating to the skin, eyes and mucous membranes.

A recent Ontario Ministry of Health report indicated that one in every 4.4 workers at the Inco converter plant in Copper Cliff had chronic bronchitis. The control sample showed that in Ottawa only one in 12 had the same affliction.[18] The suspected cause in the converter plant is sulphur dioxide. Professor Ham recommended that those affected be eligible for compensation, but because the evidence is not totally conclusive, the WCB will not recognize chronic bronchitis as an industrial disease.*

The best known respiratory disease among miners is probably silicosis. It is caused by the inhalation of silica dust which is produced as ore is broken up. Widely feared in mining communities because of its prevalence and seriousness, this disease reduces the capacity of the lungs to absorb oxygen and makes breathing laboured. "Ontario ores of nickel in the Sudbury Basin and some ores of iron, contain up to 10 percent free silica. Gold and copper ores contain 15 to 35 percent... and the uranium ores in the Elliot Lake area 60 to 70 percent."[19]

The Ontario government does not regulate dust levels for mines, instead relying on "guidelines" set by MAPAO.[20] The companies are under no legal obligation to accept these standards, which the unions, in any case, have termed worthless.[21]

Like many other industries, mining relies extensively on diesel and other combustion engines. The fumes emitted by such machines pose grave problems for those working in enclosed, underground spaces. Among the compounds released in the breakdown of diesel fuels are benzene, benzopyrene, and chrysene. Benzene poisoning produces dizziness, loss of appetite, irritability, and headaches. More important, it can make people susceptible to leukemia. Benzopyrene has been found to produce cancer of the skin, lungs, liver, and kidneys in nine different species of animals. Although proof of its carcinogenicity in humans is not yet conclusive, tests with mouse and human tissues have shown an identical response to this substance. Chrysene has also been shown to produce cancer in animals, but no reports on humans are available.[22]

In some cases, catalytic mufflers are fitted to exhaust systems of diesel machines, but these are not always kept in good repair. The more realistic solution to the problem of emissions seems to be the

---

*Generally, workers exposed to irritants such as sulphur dioxide but also acid fumes and welding fumes stand a far greater than average chance of contracting bronchitis.

introduction of electrical machinery to replace all diesels. But this is an expensive process and many corporations, including Inco, stay with the diesels. The United Steelworkers assert that the only way to get the companies to institute this solution is through legislation.[23]

> *West Virginia miners have won changes in state safety legislation to ban diesel-powered equipment underground.*
> — *The Miners' Voice*[24]

A variety of machines also produce oil mists which, volatilized through heat, are extremely hazardous. These mists can cause dermatitis, conjunctivitis, stomach disorders, pneumonia, and again, may cause cancer. The solvents used to degrease and clean machinery can even be harmful.[25]

Another industrial disease, largely unrecognized and wholly preventable yet incurable, is deafness. The implicit attitude of most companies — and the mining companies have not been exceptional — has been that excessive noise is simply part of some jobs.

Until 1968, MAPAO did not conduct noise surveys and it was not until 1974 that it established maximum allowable noise levels. This limit — 90 decibels experienced over an eight hour period — is still excessive, according to the Canadian Hearing Society. The threshold limit recognized by the American Conference of Government Industrial Hygienists is 85 decibels for a shift.[26] The noise of a drill bit on hard rock can reach 125 decibels and other mining operations produce levels well over 100 decibels.*

The corporate solution to dangerous noise levels is ear protection for each individual worker. Yet earmuffs and plugs are hot and bothersome, and can cut down on the wearer's ability to perceive other dangers. The only effective solution is to muffle the source of the noise or otherwise devise quieter machinery. But, in the words of the Ontario Royal Commission, there "has been little evidence of a concerted effort on the part of the mining industry to establish, in co-operation with equipment manufacturers, labour and government ... standards of noise generation for mining equipment as a whole."[27]

TWO VIEWS ON HEALTH AND SAFETY
The Royal Commission on Mine Safety crystallized two basic views on

*The decibel scale can be confusing. An increase of ten means that the sound intensity has been multiplied by ten. An increase of three is twice as intense and does twice as much damage. Ninety decibels is ten times the intensity of 80 and one hundred times that of 70.

health and safety. The first represents the traditional attitude and is exemplified by the MAPAO brief to Dr. Ham: "The industry takes the position quite simply that management is morally, legally and financially responsible for the occupational safety and health of the employee and cannot be relieved or excused from this responsibility. Further, the industry's position is that this responsibility is one that cannot be shared or delegated to a committee for this can only result in confused lines of authority and poorer safety."

Despite the apparent concern that corporations not be "excused" from their social responsibility, this argument is really just a defence of "management rights". In the face of evidence to the contrary, it maintains that the existing system of self-regulation has worked. Clearly it is a defence of the status quo — maximum corporate control, minimum union and government interference.

The real priority given to safety by mining companies was well illustrated at a coroner's inquest in October, 1976 into the death of miner Thomas Beals. The previous August, Beals had been killed while attempting to put a derailed ore car back on the tracks at the 2000 foot level of Inco's Frood Mine. Three miners testified that there had been about eight derailments per shift at the level in question and the responsible area foreman claimed that he had decided over a month previous to the accident that the tracking should be repaired.

Police photographs confirmed that the track was in bad shape at the time of the accident, yet a track repair crew had been assigned to work on the 2600 foot level, an area of higher production. "It was a decision of senior supervision of the mine," testified area foreman Kenneth Cook. The area in which Beals was killed had apparently never been inspected by the Ministry of Natural Resources engineer responsible for the mine, Balfour Thomas. In the preceding two years, Inco's own safety inspector had visited it once.[28]

The second approach to health and safety, in contrast to the company position, was put forward by Dr. Ham in his final report. He recommended the establishment of a new system in which "worker-auditors" would regularly inspect work sites and in which joint labour-management health and safety committees would share responsibility. But some union critics feel that Ham's specific recommendations for implementing such a system are too timid and that they would provide "neither the time nor the authority to do the job properly".[29]

However, the logic of this concept is consistent with a fundamental democratic principle — those affected by decisions should be the ones who make them. Mine workers are in the best position to judge risks and propose solutions. But the serious implementation of this principle must ultimately challenge the very notion of "management rights" and the right of management to make other decisions affecting the lives of their workers. The spectre of workers' control haunts the board room.

Interview with John Gagnon

INTRODUCTION
In 1939, Bradford Hill reported to Inco's British subsidiary, the Mond Nickel Company, that workers at its plant in Wales stood a far greater than average chance of contracting lung and sinus cancer. Although the risk had been reduced after conditions were improved in 1925, workers who were employed prior to that year suffered mortalities from sinus cancer at a rate 100-900 times the national (U.K.) average. They also were dying from lung cancer at a rate five to ten times the national average.

It is clear that Inco knew from this experience that nickel reduction operations could kill workers. Inco was also aware that the risks could be drastically cut if sufficient precautions were taken. Yet the conditions at its sinter plant in Sudbury (which was not set up until well after the Wales situation had been exposed) were not improved. Now that plant has been definitely linked to lung and sinus cancer among those who worked there, though it was closed in 1962. Today the Workmen's Compensation Board of Ontario recognizes the risks faced by former sinter plant workers and compensates those who have cancer and the dependents of those who have died as a result of their working in the sinter plant.

The following interview with John Gagnon, Chairman of the Sinter Plant Action Committee of Local 6500 of United Steelworkers of America, reveals the extent of the concern which Inco has for its workers. A Company employee for twenty-five years, John has been struggling to secure compensation for those who worked with him in the sinter plant. With the help of the other members of the committee, there have been many successes. However, many victims of Inco's operations still do not know that they are eligible for compensation and so the tragedy continues to be publicized by the committee.

INTERVIEW WITH JOHN GAGNON, NOVEMBER 8, 1976
Q: When did you start working in the sinter plant?
JG: 1951 ... on that very day in 1951 — it was so cold, 49 degrees below — and a busload of men in Coniston got hit by a train on the crossing. There was a number of men that died. From nine to twelve

died in that accident. Some died later. That was my first day.
Q: What were the conditions like in the sinter plant at the beginning?
JG: The first day, when I walked in the door until I got inside the office — a distance of some 400 feet — I never saw anything but the head of the guy in front of me and a gleam of light upstairs. That's how bad it was.
Q: It was thick?
JG: Just thick with steam. It was cold in the plant and there's a lot of cooling in the process that had to be done with water. As a result you got steam. Along with the dust. It was mostly dust, though. I don't know what made me stay there. I owed a couple of hundred dollars to a friend of mine and I wanted to pay him back before I took off.
Q: It seems that before they put in the sintering process, the company had had an experience in England with the thing. In 1939 they did the first study of workers in the sintering plant in Wales.
JG: So if they made a study in 1939 there must have been a reason to begin in the first place. There must have been a rash of ailments similar one to the other to justify this study. Because no one comes and starts a study on an industry that hasn't proven to be fatal in some capacity. They made the study and they used death — lung cancer death — to justify it. So (workers there) must have been dying....
Q: This is in the same process?
JG: Nickel oxide — it was the reason for this epidemic. There was nickel oxide dust. And that is what you are making here — nickel oxide. Since there was a lot of it, it should have been an indication to the company that something could have happened. And yet when we talked about it in negotiations they always defended (themselves) that there was no silicosis in this kind of dust and that therefore it was completely safe.
Q: It wasn't a question of silicosis, was it?
JG: No.... Silicosis is something that every miner is very much afraid of. But (at that time) they (miners) didn't know about anything else.
Q: In your experience working in the sinter plant, what was the company's attitude towards any danger that was present?
JG: Well, the company had a very negative attitude on that kind of stuff. We had to go to a lot of strikes — little wildcats within the plant itself — in order to get the conditions rectified, improved. There were times when there was mass nosebleeding among the workers because of the gas. It was later said that it was one of the worst gases there was. The people had to take the law into their own hands and shut everything down. When the supervision came and said, "What's going on?", then you'd have to say, "Look, we just can't work there. People are bleeding at the nose. We won't start anything before this thing is

rectified and the gas is stopped."

So in every case that we did that they started the units one at a time. Whenever they located a gas leak they fixed it up. But you really had to go after them and leave no alternative, no elbow-room for them to play. Otherwise they wouldn't bother. They just looked at production. That's all there was to it. That's all that seemed to concern them at that time.

Q: How many workers were in the plant at that time?

JG: I figure about four hundred per day.

Q: Did people stay a long time in the sinter plant or did they go in there for a couple of months and then move on because the conditions were so bad?

JG: Yes ... people tried to escape from that as they would escape from prison .... This is one of the reasons we can't locate them. Many people stayed there for a year or six months and then took off. Today these people are eligible for compensation and they are not here any longer.

Q: Is that just part of northern industry and the mining industry? People take a job and move on?

JG: No. It was worse than that. People wanted to get away because it was too dirty. They tried to get transferred to some cleaner area of the smelter or even underground. It was an escape from the sintering plant.

Q: How long do you have to have to be eligible for compensation?

JG: Six months for those who worked there prior to 1952 and a year from 1952 on.

Q: Why the division point?

JG: They claim that it was dirtier in the years prior to 1952 and it gradually got better. I don't share that opinion at all. I think that when it was really dirty people used the mask that was provided. When it became a little bit cleaner they didn't wear the mask any more.

Q: Why would people not wear the mask?

JG: It was a badly designed mask ... There was a lot of acid there and they gave you a mask that was made out of aluminum. You know how it oxidizes with the acid .... You had to really go after it to keep it clean. First thing you know you had a thing that stunk up your face so badly you didn't want to wear it at all.

* * * *

JG: When I got on the job today, I received a note from a guy that's got sinus cancer. I didn't even have him on the list (of those eligible for compensation).

Q: Is he a local guy?

JG: Yes, he still works in the industry but somehow he got compensa-

tion directly from the Workmen's Compensation Board via his doctor... He got in touch with me today and told me he had sinus cancer and that they were giving him cobalt treatment. So you know what that is? That's a case that's too far advanced ... he receives a pension of sixty dollars a month.
Q: From the Board.
JG: Yes ... it's crazy.
Q: What sort of struggle did you and the committee have to go through to get recognition of these claims?
JG: It's a long story. 1958 — there's already a couple of guys that have lung cancer, cancer of the bowels and so on. And died. My God, my theory was taking shape. I had always been telling the guys, doggone, there had to be something to that. It was too dirty. Since nature did not provide for a certain amount of nickel dust in your lungs, it was because it didn't belong there. And we were going to have it there and it was going to mature into something bad. It certainly wasn't natural as far as I was concerned.

Nels Thibault (of Mine Mill) said to me, "Johnny, don't beat your head against a wall. You'll never crack a nut like Inco."

I said, "I don't know. They got Al Capone. I might get Inco." So I went ahead. I visited the widows of people who had died and people in the hospital who were sick. When they died I tried to get an autopsy to find out the cause of death. About this time, 1960-61, it was taking shape quite a lot. I think there were six or seven who had died of lung cancer by that time. I was getting to be more and more assured that I was on the right track.... Dr. Sutherland (of the Ontario Department of Health) made an investigation of this and found out that we had a good point — that there was more than the national average of lung cancer coming out of that place.... They had the probabilities since they had knowledge of the epidemic in (Wales).... They started the investigation and it wasn't before 1967 that (the Workmen's Compensation Board) began to accept cases for compensation.
Q: This was five years after the plant closed?
JG: Yes, that's true.
Q: Why did the plant close?
JG: It closed because they had a more economic process.
Q: It wasn't for reasons of health?
JG: No, absolutely not.... the shutdown of the plant was not caused by the health of the men. Definitely not. They had a much more economical plant in the fluid bed roasters and with a third of the men they made twice the production right off the bat.
Q: Throughout this period, how did Inco react to your constant pressure?

JG: Their first reaction was that they thought I was a little bit on the nutty side . . . that I was going around grabbing things from the air to stuff my pockets. That kind of attitude. It was said to my fellow workers, "Well, don't worry about him. He's harmless. He's not violent. Just put up with him."

It didn't last very long. I used to go around telling the guys, trying to impress on them the importance of wearing their masks on the job. It wasn't viewed very well. (Inco) looked at me as an agitator.

As a result guys got scared and just took off. They didn't want that. They wanted guys to stick there who had experience. Then they started to get nasty. There were some that called me a sissy. (Other workers) said, "It's good for you — stainless," as they hit their chest.

I said, "I hope to hell you're right, because I don't view it that way." Many of them, bless their souls, are dead today. Those who used to hit their chests. Great big men, powerful. But, you see, you don't measure up to cancer. It doesn't matter what size you are or what resistance you have.

Q: How many have died as a result of exposure in the sinter plant, according to your calculations?

JG: We're up to fifty now who we assume have died. And double that amount are ill — lungs taken out, sinus cancer and such like. It's not a happy situation.

Q: Are there others who have left the area and might have developed illness and perhaps died?

JG: This is the worst thing on our shoulders . . . . We must try to locate some of these people. From our statistics, which are not very good, we have a list of five hundred people who we know were exposed. Of these fifty have died of lung cancer. This means that if the law of averages prevails there must be at least forty who have died but we haven't located. This is what is hard on me and the other members of the committee. We feel we are powerless to reach (their dependents). We are doing everything we can. We put out as much publicity as possible because we know they're out there to be reached. Many of them locally. They just went outside the industry into construction, things like that . . . .

Q: Was it hard to get lists of workers from the company? Did they co-operate with you?

JG: We asked Inco for their co-operation at the Department of Health level. They said, "Sure, no problem at all." So they sent us 175 names. My name was missing.

Q: This is what they said was the total?

JG: Yes, and I have correspondence to that effect. I wrote them a letter saying that this was ridiculous because there were four hundred working there in one day. And they sent us a list of 175 names. Surely somebody was trying to make a joke.

So they sent us a few more and a few more. But they sent us a letter first (saying) that this was it. As far as they were concerned that was the accurate list.

In a statement made to Inco officials at an interim bargaining session on November 5, 1976, John Gagnon said:
This company's constant refusal to allow special format for pensions of former sintering plant workers suggests strongly that in order to qualify for their Board of Directors, one has to leave behind all sense of humanity and compassion and that you also have to agree to act as a computer programmed to consider only things related to profit.

JG: My pet project at the present time is to adjust the pension (from Inco) according to the time you spent in that plant. We're suggesting time and a half for the time you spent there. If you spent ten years, you would be credited for fifteen years. This added amount should add to your age so that you're able to retire a little bit early because of the fact that your life expectancy is reduced for having worked in the sintering plant.... It's only a fair proposal but it won't be implemented until they accept their responsibility for this whole thing.

Q: So far they don't do that?

JG: They don't accept their responsibility. The things we are asking for should be automatically granted if they accept the fact that they are responsible for the tragedy. They have said many times at the bargaining table that they accept the fact that it was a tragedy that could have been avoided. But they are not ready to face the consequences of extra benefits to those who have been exposed to that place. They're having it from both sides of their mouth.

Q: Why do you think the Ham Commission treats this so briefly and says simply, "The identified causes of cancer from nickel operations have been removed." (P211)

JG: But they haven't. Most of those who died while working in the Fluid Bed Roasters, the plant that took over from the sintering plant, had worked in the sintering plant. How much of their lung cancer is attributable to the new plant? No one will ever know until someone who has not worked in the sintering plant, who has merely worked in the FBR, becomes ill. Since the FBR only started in 1962 it will be at least 1977 or 1980 before this starts to show. Because it takes fifteen years to emerge. How can they, with a straight face, say that there is no evidence? Sure there will be no evidence. There was no evidence of the sintering plant's effect until after so many years.

Q: That appears to be the worst aspect of a system in which people have to start to die before (the causes) are taken as given.

JG: This is what I have always said. Why do I have to die before something is done? Why do fifty people have to die? Now, we're not talking about one. One could be an accident — an oversight. Fifty people are dead. And yet there's a lot of dust not being checked in the FBR at the present time. I was there. I've been there now for the last two years. I've gotten a lot of things improved but there's still a lot to be done... But there are other areas that stand to cause epidemics. You've got the nickel oxide area of the iron ore plant — the recovery building. That was one of my greatest concerns when I presented my brief to the Ham Commission. There is no mention of it (in the final report). Yet sooner or later something's going to happen there, I'm afraid. I hope to hell I'm wrong because I don't want to be proven right at the expense of lives. If it happens it won't be because I didn't warn them.

Again — you're labelled as an agitator, an extremist, someone who is spitting up in the air. Well, do you know of a better idea? You have to argue *before* it happens, not after. Afterwards you only pick up the pieces. But if you can avoid an explosion — that's what you're after.

Yeah. I've been in hospitals seeing friends who have worked with me for ten, eleven years. One case in particular that really got me was when I went to see Johnny Welyhorsky. He died about two days after I had seen him. I got there and he said, "Johnny, I'm going to die". So I went to try to comfort him. He just grabbed me and lifted himself right out of bed. He was just light as a feather. There was nothing left of him. He was totally eaten up by cancer. And he said, "You can't do anything for me, but work like hell for the others." I never really got over that.

That was around the time Weir Reid died... I had a nervous breakdown and lost three months work. It affects you pretty bad sometimes. You're so involved that sometimes you take too much. Some of the griefs of others become your own. Nature only made you to take so much and you can't go beyond that. But I came back. Like they always say — the cat came back.

Q: Do you think you will be successful in achieving the high goals you've set for yourself, given the way the system is working?

JG: No. But you see, if you're going to achieve things you have to aim high. Perhaps I aimed a little bit too high but I'm going to work like hell to see what I can do. I was sincere when I made the aim. I swore I was going to make the company pay for every bit of grief that they have caused. Families and so on. I'm trying hard. I'm not getting anywhere too fast but I'm making some gains....

... AND BROKEN BODIES 141

JOHN GAGNON

Why Inco?

Why is it important to examine the particular history of one company? Why expose to public scrutiny that company's relations with its employees? Why question the effects of its investments around the world? Because almost from its inception Inco has been a microcosm of the larger economic system.

The stimulus for the large scale production of nickel can be traced back through technical to military, political, and ultimately economic needs. In the last quarter of the nineteenth century the quest for a way of hardening steel became intense. The immediate motive for this search was military — the need to develop artillery shells which could pierce newly-developed armour plating. This military imperative stemmed from the competition among the larger nations of Europe for colonies. In turn, this struggle was a contest for new markets and sources of raw materials. Its intensity can be attributed to the rapidly expanding productive capacities of those countries in Western Europe and North America pioneering the development of capitalist enterprise.

For Canada, the significance of nickel production extends beyond its place in the evolution of capitalism. Nickel mining here may well have been delayed for decades had it not been for the particular nature of Canadian Confederation. For not only did the building of the Canadian Pacific Railway result in the recognition of the value of the nickel-copper deposits of the Sudbury Basin, it also provided an efficient bulk transportation system which allowed for their easy exploitation. Rather than being mere coincidence, this order of development is typical of the brief history of Canada as a nation-state. Canadian entrepreneurs have been most successful organizing sophisticated transportation and support systems necessary for extracting raw or semi-processed natural resources for export from the country.

While the founding of Inco's ancestor, the Canadian Copper Company, illustrates so well the interplay of economic, political, and technical forces, there was no one compelling reason why it succeeded and defeated its competitors. It enjoyed a headstart in securing mining patents or claims and one of its founders, Samuel Ritchie, had the knowledge and foresight necessary to pursue the military potential of nickel. But he was deposed from his own company and his commanding position assumed by Colonel R.M. Thompson. Ritchie, despite his skills, was unable to dislodge from its primary position in nickel mining the company which he had founded.

The fate of the nickel combination under Colonel Thompson further illustrates the impersonal forces behind corporate growth. After the Canadian Copper-Orford group had attained the basic requirements for economic success — a source of supply and a market — it became subject to the almost irresistible pressures which led to the formation of International Nickel. The instruments for this process were the same financial interests which had been behind the monopolization of the steel industry in the United States (represented by J.P. Morgan) and the corporate lawyers who had acted for the Rockefellers in their consolidation of the Standard Oil monopoly (the firm of Sullivan and Cromwell).

The creation of the U.S. Steel and Standard Oil monopolies was complicated by the relative maturity of both the industries involved. Not only were there a number of producers in each case, making it difficult to form a total monopoly, but the centrality of both steel and oil to so many other important industries meant that there would be considerable opposition to their monopolization. This resistance eventually forced the partial dismantling of both trusts and the development of slightly less centralized oligopolies.* But in Canada, International Nickel, the handiwork of these same manipulators, survived virtually unchallenged for a half century. For nickel was being produced in North America by a very few companies with one consortium overwhelmingly dominant and the impetus for the formation of a single producer came from the prime users of the metal, the steel fabricators themselves. There was little difficulty forming a

---

*In an oligopoly, two or more companies together dominate a market and set prices, usually by one of them assuming the role of price setter. The whole question of "trust-busting" in the U.S. during the early part of this century is shrouded in confusion. Though liberal historians treat the apparent breaking of the Standard Oil and U.S. Steel monopolies as a triumph of free enterprise and democracy, others suggest that it was the triumph of one group of industrialists and financiers over others, and even at that, a largely illusory triumph.

"monopoly" and no reason for resistance from the significant customers.*

Throughout these early years of development certain questions about nickel mining's value and to whom it would accrue did arise in various forms. But Canada's political leaders from John A. MacDonald to Wilfrid Laurier and his successors, were contented with minimal benefits — the jobs created in mining and the odd personal pay-off. Those who argued for governmental action to force the further processing and fabricating of nickel in Canada were rebuffed, whether or not they argued from their own personal interests as in the case of Samuel Ritchie. Government leaders accepted and repeated as their own the position put to them by American corporate managers — development on the terms dictated by the corporations or no development at all.

From the beginning, those closest to the mines, the area farmers and the Company's workers, had the most compelling reasons to resist the corporate will. It did not take long for the mine complex's pollution to affect local crops. The persistence of the farmers in their protests did result finally in official recognition of their problem but the fruits of their dissent were ultimately paltry.

For those who daily worked and lived under the Company's regime, resistance seemed even more futile. Even Inco's trading with Germany during World War I was not enough of an affront to official morality to prepare the way for the unionization of the workforce. For this latter was a polyglot mix of immigrants and French Canadians, convenient targets for wartime antagonisms. The only real consequence of the Company's questionable war record was the passing of tax legislation in Ontario which effectively forced the construction of a refinery in the province.

War's end and the rapid reduction of military budgets plunged the nickel trust into a brief period of instability. But by the mid-Twenties, International Nickel had acquired a new management which vigorously sought new markets for its product, particularly among consumer goods producers. Near the end of that decade Inco consolidated its monopoly by merging with the Mond Company and moving its legal residence to Canada.

The Depression initially affected nickel production as adversely as it did most other industries. But Inco recovered rapidly, aided by the re-militarization of Europe which accelerated with the passing of the decade. World War II prompted the enormous expansion of the North American economies. As the one secure supplier of nickel for all the Allies, Inco's fortunes soared, too.

*As noted previously, this was technically the formation of a monopsony.

But war also improved the fortunes of the average worker, for modern warfare required mass participation. Neither foot soldier nor labourer on the "home front" could be denied their own importance any longer. In Sudbury, the nickel workers, roused by the Company's crude repression, finally succeeded in forming a union.

In peacetime, much of the certainty of war vanished. Empires had fallen and others were rising. Wartime allies became peacetime foes. Politicians and diplomats needed to justify the new and sometimes violent division of the spoils. Both governments and individuals ran afoul of the evolving realignment.

In Sudbury, the local union resisted the new orthodoxy and continued to tolerate Communists within its ranks. Old rivalries revived and new ones surfaced. Suffering official isolation from most other unions and the lack of a broad working class political movement, Mine Mill slowly succumbed to the climate of the times.* Its leaders committed errors common to other unions but without the protection given through affiliation to the official labour movement, their local became prone to raiding. And many of those harbouring grievances against the leadership or entertaining their own ambitions gradually learned to express themselves in the language of anti-Communism in order to gain attention and support. But should that sort of ideological ruse seem unusual in a world then divided, so it had been declared, by an Iron Curtain?

Little countries, too, like Guatemala, found themselves curiously out of step with the march of events. Depression followed by war on an unprecedented scale had seemed to usher in a new era. For was not the United States, the giant emerging full grown from battle, proclaiming an end to colonies and presumably to empires? Unfortunately the giant had its own needs, every bit as insistent as those of its decrepit European rivals. And as these needs came into conflict with the reforms attempted in Guatemala and in other countries around the world, the American banner of democracy was replaced with that of anti-Communism. Strangely, the latter often seemed to demand the suppression of the former.

In 1954, Guatemala was turned onto the path of righteousness by the United States and its friends. Indonesia did not succumb to Ameri-

---

*Though areas of working class numerical dominance like Sudbury may have been in the early stages of developing the popular political culture necessary for a mass radical movement, that development was quickly truncated by the antagonism between the CCF and those in Mine Mill identified with the CP. In any case, such a culture could not have developed in isolation from the rest of North America.

can pressure until 1965.* Neither Arbenz nor Sukarno wanted to present a fundamental challenge to the American economic model. They merely wanted to modify it in order to implement it in their own countries. Naive to some degree, each in his own way tried to off-set by using state power the disadvantages their countries faced as "underdeveloped" in competition with the well-developed. The threat made by the American and European transnationals operating in these countries, which was supported by their governments, should be familiar to Canadians — development by the transnationals or no development at all. In practice, "no development" meant not just the kind of external pressures that powerful corporations and governments can exert on weak nations, but also direct interference in the internal politics of the latter.

Where does Inco fit in this maze of international politics and intrigue? It did not appear to have any immediate goals to attain through the conduct of American foreign policy during the early years of the Cold War.† Yet, its later investments in Guatemala and Indonesia show how the defence of the general interests of the larger political-economic system protected future investment opportunities for particular corporations. It should also be obvious that the kind of economic development sanctioned for the Third World by American might and implemented largely by transnational corporations like Inco can often be enforced only by the most ruthless and brutal authoritarian regimes.‡

After World War II, little public debate took place on the type of economic strategy to be followed in Canada. There seemed to be no alternative to the American Way. The Cold War ensured the silence or silencing of many of those who might have looked for other ways. The

*Iran experienced a similar intervention in 1954, too. From 1946-1949, Greece had been the recipient of more overt aid, given by both Britain and the U.S. in what was to be the precursor of many interventions around the world over the next three decades.

Coincidentally, 1965 was also the year in which American marines landed on the shore of the Dominican Republic to prevent the reclaiming of its presidency by Juan Bosch, who had been elected but deposed by the mitutary dictator Donald Reid Cabral in 1963.

There should be no need to mention the more recent and widely-discussed interventions by the United States in Indochina or in Chile.

†As it happened, longtime Inco counsel and director John Foster Dulles did play a very direct role as U.S. Secretary of State, in Guatemala and to a lesser extent Indonesia.

‡Not all Third World countries host to American corporations are dictatorships. But those often offered as models to more reluctant governments are, eg. Brazil, Iran, South Korea, and Indonesia.

immense natural wealth of this country and its small population also allowed a standard of living high enough to discourage any fundamental and potentially divisive questioning. But in recent years, as the economic future became less certain, the terms under which mining companies in particular operate in Canada began to be scrutinized.

In Manitoba, a New Democratic government commissioned economist and former Liberal federal cabinet minister, Eric Kierans, to analyze the role of mining in that province's economy. He concluded that the mining companies operating there, of which Inco is the largest, were making superprofits as a result of the relative scarcity of the ores being mined.* These profits, in excess of the normal return on investment, are called economic rent. Logically, this rent should go to the owners of the resource who in Canada are the people as represented by the provincial governments. Kierans argued that the provincial government should appropriate the surplus by taking over the mines (with compensation to the present operators for their investment).

The response of the mining lobby to the so-called Kierans' Report was predictably threatening. Government intervention of the sort proposed was equated with socialism. In fact, Kierans is a national capitalist. With the revenues to be generated by provincially-owned mines, he would lower taxes or in some other way provide incentives to develop a strong private manufacturing sector. For, he warns, without the nurturing of more secondary industry in Canada, the depletion of the country's natural resources will lead to a drastic lowering of living standards here.† Manitoba's social democratic government was not prepared to implement anything so controversial as Kierans' plan.

Inco Ltd. is the world's largest producer of nickel, a strategic industrial base metal. The company is Canada's largest mining company and is the country's leading producer of copper as well as nickel. It also produces gold, silver, iron ore, platinum, cobalt, sulphur, selenium, and telerium.

Refining is carried out in Canada and the UK and the main rolling mills are in the United States and the UK. Inco has its most important markets in the United States, Europe and Japan.

Inco also controls ESB Ltd., the largest battery manufacturer in the

---

*This is scarcity relative to demand. Kierans reached his conclusion by comparing profit rates for companies involved in the separate aspects of production, ie. those only mining, those only refining, and those doing both. He concluded that the superprofits were being gleaned at the mining stage.

†Reaching a similar conclusion, a study for the Science Council of Canada warned that "before the children of today could reach middle age most of the resources would be gone leaving Canada with a resource-based economy and no resources," if present policies are followed.[1]

U.S. and Daniel Doncaster & Sons, a British metals fabricator. It is well-entrenched in Japan and owns 10 percent of Impala Platinum, a mining company in South Africa.

Lateritic mines, coming on stream in 1977, are located in Indonesia and Guatemala and Inco is pursuing a promising deposit in the State of Goias, Brazil. The Company is also heading up an international consortium which is developing sea mining techniques. Other exploration is proceeding in Australia, Canada, the U.S., the Philippines, Guatemala, and Mexico.

As of December 31, 1976, Inco's total assets were $3,628,311,000. In that year, its net earnings were $196,758,000 on net sales of $2,040,282,000. It delivered 409,830,000 pounds of nickel and 355,990,000 pounds of copper. Over thirty-eight thousand of the Company's employees were involved in its metals business; another 17,000 worked for ESB, its battery subsidiary.

Not only at the level of international and national political affairs did the Cold War help private enterprise. The resistance of workers on the job was often weakened by the ideological confusion it fostered within the labour movement. The failure and aftermath of the Mine Mill strike against Inco's Sudbury operations in 1958 is a good example. Rather than analyzing the reasons for failure, the local's members became involved in an acrimonius political dispute. Not until the late Sixties had the Inco workforce in Sudbury achieved once again the unity necessary to confront the Company.

Though that unity has been most visibly tested during contract negotiations and strikes, it is probed daily by the Company's management for signs of weakness. For the success of a company like Inco depends on the ability of its corporate leaders to manage their resources, including the humans whose labour power they purchase. And management is judged not by standards established within a single company but rather according to criteria imposed by the larger economic system.

In the final analysis, a corporation or any private enterprise must be profitable. Its ability to borrow in order to make new investments and indeed, its ability to retain shareholders, depend on its profit performance. Failure to attain levels comparable to other firms will pose immediate problems for the highest strata of a company's management. If a new business strategy is to be attempted, it will be passed down through the bureaucratic hierarchy which all large corporations require. At each level the plan will be adapted and made more specific in an effort to reconcile the original goals with the reality of the work situation. Ultimately, the larger demands placed by the economy on the corporation will have been translated into demands on those at the bottom of the corporate pyramid — the workers.* Though this process becomes more obvious when a company does face a particular management crisis, as Inco did during the first years of this decade, it is typically a less dramatic, continuous operation.

Those who train for, or aspire to management positions in industry must learn to equate efficiency with maximum profitability. But most production workers are more likely to be more sensitive to their own and their co-workers' everyday human needs. A collision of interests is inevitable. The resulting friction takes many forms. An individual worker disobeys a supervisor who tries to discipline the offender. Efficiency experts continually seek new techniques to increase productivity; a group of workers refuses to change established work procedures. Labour relations specialists, psychologists, lawyers, and economists work to get union acceptance of the contract most advantageous to the company; the local rejects the "final offer" and its members go out on strike. On another level, a national union may conduct a campaign to get improved safety standards enforced by government, while the opposing industry association may argue that the existing standards are adequate or perhaps even too stringent.

Neither is this a game nor a contest dependent on the individual participants. Union pressure may force the replacement of a tough supervisor with a more skilled manipulator. A new work quota may be revoked, its desired effect to be attempted through some other means. A strike may be won but the company may decide to re-locate in a lower wage area. For the logic of the system remains and persists.

A company like Inco will always try to place the responsibility for accidents on the individual workers, while denying the need for the collective participation of all the workers in alleviating safety problems. It will always deny the dangers to health of its pollution or at least shift responsibility for correcting such abuses to government. It will always meet attempts to regulate it through taxation with threats to move to more hospitable locations. (In the words of Inco's chairman, J.E. Carter, only "taxation and other restrictions imposed by Canadian government" can make Canadian nickel "uncompetitive".)

So why attack Inco if it is only a part of a larger system? Noranda operates not just in Canada but now also in Chile. Hudson Bay Mining and Smelting, with a large mining complex at Flin Flon, is only a small

---

*In actual fact, no corporation operates as single-mindedly as suggested here. Internal bureaucratic politics will modify any plan implemented in a corporation. The appearance of efficiency or competency is every bit as important to the ambitious executive as the real thing, though at the highest level somebody has to accept responsibility for the balance sheet.

part of the Anglo-DeBeers empire of South Africa. The role played by Alcan in Guyana and Jamaica is questionable, as is that of little Sherritt-Gordon, phasing out one town in Northern Manitoba even while it collaborates with a social democratic government in building another. And what about the giants — Canadian and foreign-based — in other industries, the Stelcos, the Eatons, the Royal Banks?

All their stories would highlight different points, expose various aspects of the larger system, reveal particular ironies and injustices. For the sake of those whose lives are affected by these companies, their stories deserve to be told. For the sake of those who work every day for Inco, who live under the path of its pollution, who have died for it or who have been killed so that it and corporations like it could operate in Guatemala and Indonesia, for them this story is told. But not just for them.

For if each abuse by each company is but the result of the inexorable logic of the larger system, then the old trade union slogan "an injury to one is an injury to all", must be given new meaning. And an attack on one corporation, be it Inco or any other, must become an attack on all, on the entire corporate capitalist system.

Appendix
Working in Thompson

The following interview took place in the spring of 1974 in Thompson, Manitoba. The participants were Ashley Chester, Harold Chorney, and Peter Richards, all of whom were involved in a video education project, and Lloyd H., a young worker with considerable experience working in various parts of the Inco complex in that city 500 miles north of Winnipeg. The interview has been edited in order to be presented as a straight narrative.

Without doubt, some aspects of life working for the "Big Nickel" have since changed, particularly with the deteriorating economic climate in Canada. However, Lloyd still seems to have captured the essence of work in many large industrial corporations and has expressed it with good humour. Hopefully some of that verve will come through in this transcript.

Shortly after this interview, Lloyd returned to his native Saskatchewan to study for a certificate in technical education. At present he is a member of a millwrights union.

* * * * *

I came out of a nice, conservative home, I didn't drink. I didn't smoke. I came into the labour movement and I had big, bad ideas that if you didn't go to the meetings a man with a big, black bomb was going to come to your door. Oh, yeah. I didn't know what to do. They handed me my union card and I started going out and getting involved and trying to join — a good, honest union member. I tried working and working and working and working. Maybe I was working along the wrong lines. When I got to be a steward I figured I was quite privileged. So I started to help the men here and there. Lots of them still come to me to phone the office. And you can pretty well tell what they're going to say anyway. So you let them down a little or brace them or bring their hopes up. Very few grievances are won — very few. . . .

* * * * *

I can tell you, half the goddamn shift bosses up there, you know what they did? They got "qualified". You know, this is the big thing. Like Inco — every big company — what do they do when they're looking for a shift boss? They look around for the sucks, eh. Who's nipping who's bum. And then they check your record, your papers, and if you haven't blown any shifts and you haven't raised any shit, well then, you're OK, eh? And then they look at what you're qualified for. You're going to drive the motor up and down in the drift — you're a qualified motorman and you've got your name in the book. You walk in front of a train a couple of times — you're a switchman and you've got your name in the book. You drill a couple of holes, but maybe stay away from drilling because that's hard work. Maybe drive the scoop around and all this stuff. They nip, nip, nip. And they get away with it, that's the way it works. You're goddamn right.

Say I wanted to be a foreman. Well, the first thing I do is start sucking around. Like, I'd go into the lunch room and sit there with the shift boss and offer him sandwiches. This is actually true. Oh, yeah! I'm not shitting you. You're allowed half an hour for lunch. So you come in and you look at your watch, just to make sure the shift boss sitting there sees you. Sit down. You've got three or four men with you. You're always the first man that stands up and says, "Well, fellows, let's go." And everybody's looking at you like you were some kind of a nut or something.

You can pick these guys out, these potential shift bosses. You can see them come in. They're from all walks of life. But they've usually been salesmen or something. . . .

What kind of life does a foreman have, really? You've got to chum around with the other foremen, I guess. Because all the guys really despise you. So you hang around with other foremen. It's like the life of a policeman, I guess. They pay you a few extra dollars so you can go to different clubs and stuff. So you can drink stuff mixed with your rye instead of water. Stuff like this, you know. . . .

The reason they haven't got absenteeism with the staff is that the staff is so damned scared of losing their job. I'm telling you they just don't dare miss a shift. You have to take a look at the type of person who's going to get into that position anyway. If you looked at their record, the people on staff back when they were hourly, you'll find out they didn't miss any shifts there either. . . . It helps if your daddy was a big wheel in town or a general foreman or something. This certainly helps quite a bit, especially with Inco. It's quite handy if you've got a big dad in Sudbury.

* * * * *

I'm lazy, and proud of it. If you go up to your foreman and he says, "Are you a hard worker?", I say, "No, I'm the laziest son of a bitch that ever came along." And he says, "Oh." What's he got to work with now? You look at the position you're leaving him in. He's totally, abscondedly, screwed.

I actually did this when I was working in the smelter, because I hated the smelter anyway. What do they do to you? They call you to the office and give you a big talk and say, "You're not happy here?" And you tell them, "No, I'm not happy here." "Would you like to transfer somewhere else?" "Yes!" And then they send you somewhere else. That's why I travelled all around the plant.

I decided that every time I saw a whitehat coming I would sit down instead of standing up and starting to work. You're working away and you see him coming. You sit down, you know. And he comes up and catches you all the time. "Well, I told you, I'm doing nothing." He looks around and sees you did a little bit and then he won't say too much. This went on for about three months. "Aha," he says, "I gotcha working!" Well, it was just like little kids with a new toy. He's just totally stupid. It's really crazy. Those damned fools.

There's only two ways to go. You've got to be a high baller — that's the guy who gives it shit. Or else you're a D.F.'er — that's the guy who takes it real cool and does nothing. You get respect both ways. But if you're in the middle, well, you're just an ordinary Joe. You can get respect if you do nothing. You get credit for it. You don't have to do anything, you just have to have good lies ready. And they promote it. Because the foreman has to have good lies to pass on to his foreman. It's got to go all the way up the line. So they come down to you and say, "Why didn't you make your breast today?" (And you tell them), "Oh, bad. Listen, just terrible. Just scaled all day, terrible." They go ahead and tell this to the supervisor, and the supervisor tells it to the general foreman, and the general foreman tells it to the mine captain and it gets lost in the dirt all the way up the line.

The foreman doesn't really come up to you and give you shit. He doesn't come out and call you an absolute, blithering idiot, the way he did ten years ago — "You have no brains, you're stupid." Because the people who are coming here now ... so many younger people are coming to this town. And they just don't put up with this horseshit. If a foreman comes up and calls him a blithering idiot, he'll turn around and say, "OK, if I'm a blithering idiot, why did you hire me?" They just tell him jobs are two bits a dozen. They just don't take it.

And the old-timers are listening to this. "If you don't have to take it, why should I?"

"Well, in Sudbury," (this is what Inco's big cry is), "We don't have these problems in Sudbury." Well, they've been brainwashing that place for fifty years. And the young people in this town know what to expect when they come to Inco. They expect to get downtrodden and everything because they hear it all the time at home.

There's so much horseshit that I've gone through with that Company. It's just impossible to put into words. Like the personnel manager, when I told him I was going to quit. "You didn't enjoy working here?" he says as he hands me a cheque for a thousand dollars and my papers.* "What are you going to do with it?" And I say, "Quit." He gives me the hint — "We've educated you for three years and now you're going to quit. Now isn't this a great place to work?" They come out with the ultimate question: "What's wrong with it?" It's like asking what's wrong with being in jail! . . . That's what I told him.

He mumbles and bumbles around. Then they go into this question — why — they are honestly trying to find the one reason why their people are quitting. (It's the same thing as the Compensation Board looking for one reason why men have accidents. You have accidents because men fall off the ladder. You have accidents because a scaling bar gets stuck through you, or a rock falls on your head. That's why you have an accident. You're just standing in the wrong spot, or you shouldn't have been using that bar, or you should have just blown that day.)

So he says, "We've got fifty different men that quit. High seniority men." And I've got fifty different reasons why they quit. If they did the same thing when the men quit as they do when they have an accident — typical of Inco, whenever a man gets hurt they make a rule against it so they're covered with the Compensation Board. They go through their ninety-nine thousand pages of sheets and it says, "Don't step on flat rocks." And a man stepped on a flat rock. See, he broke a rule, that's why he was hurt. Simple. So if they did the same thing with quits — "A man quit because a shift boss said this." OK, all shift bosses don't say that. Eventually they're going to get men to stay!

The Company had a big meeting to find out the one reason why everyone was quitting in Thompson. They called all their general foremen and superintendants and the odd shift boss who wanted to

---

*H. is referring to the bonus and certification which he received after completing an apprenticeship programme.

come. They called them all into this big meeting hall. The chairman's up there and they say, "OK, now we're taking suggestions as to why the men are quitting in the different departments." One stood up — this one, actually, he's a strikebreaker. That's how he got to be a general foreman. He was strikebreaking for Sudbury, so they shipped his ass over here — so he stood up and he said, "This recent raise we got. Every man got $32 a month cost of living bonus. On staff, we got $64 a month. I don't know about you, Mr. Chairman, but when I go out to buy a loaf of bread, it costs me the same price as the guy who works for me."

And the guy up there says, "Of course, you know the staff are supposed to eat a little better, to live a little better than the average working man." And the next guy over there, he said, "Well, maybe that's why your men are quitting."

* * * * *

The workers have initiative, too. All it takes is a serious situation for them to act. Sometimes something happens on the job where the men just naturally band together. Once you've got a band like that it just can't be broken. Once they band together they stay that way because they realize they've got to stay that way. Once they've pulled a stunt on the Company, they do stay together. Say at the Birchtree Mine the road gets snowed in and they just can't get out. So they just camp in the mine and they say, "We're staying here and we're getting paid for it." So the Company supervisors say, "OK, if you're getting paid, you've got to work" And they say, "Bullshit." So the Company ended up paying the guys who slept and the guys who worked the same thing. . . .

The situations that come up are usually accidental. In most of these situations there's been a steward involved. Sometimes it's a steward who you'd have least expected to do it — a quiet guy who sits at the meetings and hardly says a thing. One time it was a guy who always talked on the Company's behalf. He started a thing in the smelter when we all walked off because of the gas. He was even temporary foreman for awhile. I was surprised as hell when I heard that (he had led the walkout).

* * * * *

High power salesmanship gets people up here. You've seen the ads for the glorious life and all this crap. They get the men up here.

I was down on 1300 last week when they had all the new hires down there. Well, they kind of rushed the new hires in, put a hat and light on them. Because they don't want all the guys talking to them.

They spook them — "You're crazy buddy," — this is the way the guys talk to them. They actually feel sorry for somebody coming to work here. They try to scare them off.

I was sitting down there and this foreman was telling these guys — this is no shit, I was sitting there having a cup of tea — he says, "Were you talking to them?" And he gets all excited. And I said, "No, I was just asking them where they were from." And I asked them what kind of pitch (the Company) was giving.

The first thing they talk about to the new guys is safety. "Don't stand in a drift because a train will run over you," and all this crap. They tell you what not to do. To turn your light on, to turn your light off. And then they start into, "All our people wear safety glasses. We have no eye accidents, there's none at all. And in every accident, the man couldn't have been wearing his glasses, otherwise he wouldn't have had an accident." And then they say, "If you do have an accident, the procedure is to report it to your foreman," — who could be on three different levels. And if you have a broken arm you have to climb the ladders and look for him. You've got to inform him. And then he's supposed to call the cage. Then you inform First Aid. And First Aid decides whether or not you should go to see a doctor. And if they decide wrong you end up with a mark on your face. Like this, see. What everybody calls a pencil mark. They just put a little stitch there where it should have had several stitches. But, oh well, that's good enough for H. Piss on it. They're actually the ones that make the decision.

The foreman turns around and says, "Of course you can go to the doctor if you want, but you've got to go to First Aid. You've got to do all the proper things first. Otherwise it just makes it hard for everybody all around." He hints that "If you make it hard for us, we'll make it hard for you."

If you pick up an application for any mine — I don't think it's just Inco, (also) the one I filled out for Allen Potash — it says on the bottom, "Have you ever drawn any compensation?"

See, they believe there's such a thing as chronic compensation drawers, or something. If a man draws compensation, already he's blackmarked in the industry. If you've a back injury, you might as well forget it. Right away they don't believe you. They set you up for a couple of days screwing in light bulbs 'til you're half batty and you ask to go back underground. "Sign here and away you go."

* * * * *

A lot of people come here to make a stake to go back home. Most people who come to these mine towns are like homing pigeons. They've got an instinct. They'd like to go back where they came from. That's the way most Canadians are. Maybe they go away to school here, go away to work there — but they'd like to go back to where they were originally brought up. Most people are like this. And especially those that come from other countries. They come here and they can make a good wage for a good day's work. They put their money aside and when they get enough they can go back to their country and they can take it easy.

Very few people say it's nice to work for Inco. A lot of people say, "I like the money." But very few say the working conditions are good. A lot would say, "I've had worse." Or, "Could be worse." Or, "It's cold in winter but not so bad in summer." But they wouldn't come out and say, "Boy, I love working here." I don't think the men, really, as a whole enjoy working here. Or living up north. It's a good place to make money.

The personal views reflect this when anything happens in the town. Say a lawyer gives a guy a bum rap, doesn't protect him right and the guy goes to the can. The word goes around the plant — "That lawyer's no goddamn good. If he was any good he wouldn't be up here." This is the same with doctors or anything else. What they're doing is reflecting their own, personal opinion. Because most people up here kind of get that idea after awhile. It's like being in the army. Inco looks after you. They fix your teeth. They put glasses on your head when you go blind from underground. They do all kinds of nice things for you. They give you a wage that you can live on. You won't get rich on it, but you can live on it. And you work regularly, you get a regular paycheque. And after five or seven years, to actually pull out and leave town, it's damn hard. I found how darn hard myself. Because you've got to sell all your old junk and stuff because it's too expensive to haul. You know, it's a real hassle to say, "I quit."

* * * * *

The majority of people just take the union for granted. The ten dollars comes off their cheque and there's nothing they can do about it. They feel the same way about the union and its activities, there's nothing they can do about it. If they have any trouble on the job they go to the union. Usually as a last resort they go to the union and try to get things squared. Most of the people are impassive. They just walk by it and that's it. There's a few that try to fight it. There's a few that go along with it. Different people fit in different slots. There's a few that are in the union movement for the benefit of being in the movement. There's people in the union movement trying to change it. There's people outside trying to change it. There's people in it trying to keep it like it is. It's a three or four way fight.

I think the men have to get together — not really their political beliefs — but they should get together on where the labour force is going and what it's doing. If they could all get together without a union meeting and then just say, "OK, our contract's coming up. What do you think we should do? What's bugging the boys most on the job?" You might get four thousand different ideas. And then you group it down into departments and you try to find out the main thing that's bugging this department. Try to keep away from the money things even. I'm sure it would come out in the wash a lot better. Then when the men are working under the contract, they say, "OK, this is what we ask for and if we didn't ask for something else, it's on our shoulders."

The same thing that happens in the town. The men come together a lot of times, whether they're going to do something to the trailer camp or tear up somebody's sidewalk or knock down somebody's home. Everyone comes together outside of the union and outside of any other thing. Just men coming together over a common issue. They start working and get to know each other. And then something happens in the plant and the men glue together there. I think this is the way it's got to start. I don't think it can work through your traditional unions or through political pressure or through company relations or anything like that.

* * * * *

The Company doesn't hold (being a shop steward) against you, really. Because they have no fear of stewards. Why should they put you down for it? The Company looks at it this way — if you're a steward, you're just a man trying to get ahead. What's wrong with that? Unless you go around calling them a bunch of pricks and yelling and screaming at them. Naturally, they're going to put you down. Because you're yelling and screaming. But not because you're a steward. If they have too much trouble with a steward, if he's putting in too many grievances, they phone the union and tell them to quieten him down. You believe that? I went on one of these delegations. (This one steward) was raising too much shit with the Company and they got word to the executive and the executive kind of got word back to the steward's council and they picked two stewards to talk to him and tell him he should work through the office more . . . in other words, "Shut up." Because (if you) go down to the office and tell them your problems, they won't do anything about them. "Don't raise too much shit on the job — it gives the union a bad name." The union's supposed to work in harmony with the Company. They're not supposed to buck them all the time.

APPENDIX: WORKING IN THOMPSON 161

I think the steward is kind of a sinking boat. I wouldn't like to say that much. He's got more support from the men than any other representative of the union. The men tend to lean toward the stewards a little more because they work with him and he knows the problem first hand. If they have to go down to the union hall and explain it to someone who works in the mill and they work in the smelter or the guy doesn't speak too good English — he can't get across what his problem is. Whereas you can tell the steward and he goes down there. The men come to the stewards all the time with their problems. But lots of times the steward takes the problem down and once it's in the office, it's pretty well out of the steward's hands.

* * * * * *

I'd say the wives, they don't know what goes on in the union. Come contract time, I think the wives have a heck of a big say in what goes on. In how the vote turns out. The housewives have a lot of pull in whether a contract is put through. They get on the radio show and you hear on the radio, "This is the best contract. It's great. Lots of money. Certain people are going to get a dollar an hour more." Fifty people get a dollar an hour and the other three thousand get about eighty cents. A few guys get nothing.

I don't think there's any movement led to integrate the wives into the movement. Or into the programmes at all. The union's kind of a men-only world because everybody swears at union meetings. Well, women are working in the plant. Five years ago you'd say, "Women working in the plant. It would never work!" Now they're there doing as much as we do. So what's the difference, really . . .?

When you get women working in the plant or in some of these dirty, shitty places we go to work in . . . they say, "Clean it up, I'm not working in there." Whereas the men will go in there because, "I'm the primary bread winner." And they'll go right on digging in there. They'll do the dirty jobs for the women who'll go to do another job. (The women) are a bit more strong-willed in a situation like that.

* * * * * *

What is the definition of illegal? We signed a contract for three years. That guarantees the Company "working stability", or whatever they want to call it — big words like marmalade and peanut butter. What that guarantees us is that they're going to stay here and have no lockouts for three years. But let's say the power goes out in T1 mine and some goofy shift boss sends the cage through to the head frame. Which is an example. And then all the miners are off for two days on account of this, because they've got to cut it out of there and put in a new cage before they can go down. These men are sent home without two days pay. That's just tough because the cage went through the head frame. What do you call that? Do you call that a strike or a lockout? The Company's done it. It's a lockout. They're sending you home for two days because they blew it. Now what happens if the men are in the smelter and it gets so goddamn gassy in there you can't breathe. So they go and get a doctor's slip. They're going to sue the union for something like that. I think it's the same cup of tea both ways. If the Company can do it, why the hell can't we?

Afterword

Sudbury has often been dubbed, "The Nickel Capital of the World" by the chronic boosters in the local Chamber of Commerce. However, in the aftermath of Inco's announcement that 2200 jobs were going down the drain at Sudbury (along with others at Port Colborne and Thompson), it was left to the ever-reliable *Sudbury Star* to emerge, virtually alone, in its defence of the Company. The Chamber of Commerce, once a reliable Inco mouthpiece, joined the angry chorus denouncing the Company's move, which was announced on October 20, 1977. People pointed to the $380 million in deferred taxes which Inco owed, to the $70 million in EDC loans and the expansion into Indonesia and Guatemala, to the continued profitability in the face of recession and poor markets, and, not least, to the plight of the workers whose livelihoods were to disappear.

Predictably, Inco defended itself on the grounds that adverse market conditions had forced its hand. It had carried the community long enough through the latest hard times. It was a matter of economics, supply and demand, and so on. Yet, Walter Curlook, a senior vice-president, in a rare moment of corporate candour, pointed out to the *Toronto Star* that, "I don't think there's any doubt in our minds that Third World countries like Indonesia and Guatemala are much more likely to act quickly against Inco if we took measures that would seriously affect their social and economic development programmes."

It is clear that Inco's expansion and diversification programmes had a lot to do with the layoffs. Transnational capital has the advantage of being flexible and can move around the globe when need be. Inco will doubtless be feeling more secure than it did in 1971, another recession year, when there was no ESB, no Exmibal and no P.T. International Nickel Indonesia. Feeling less secure in the years to come will be the Inco workers who, in the 1970-71 recession, were working under a contract which had given them substantial wage increases.

The whole issue of such layoffs brings into question the validity of an economic system which allows a company to export the surplus generated by its workers in order to ensure its future profitability and which then allows it to eliminate the jobs of those workers when an economic downturn occurs. When the demand for nationalization without compensation was raised at a Sudbury rally protesting the layoffs, it was greeted with sustained applause by the assembled mineworkers.

Notes

Preface

[1] Chambers, W.G. and Reid, J.S., "Changing World is a Challenge for Government and Industry", *Northern Miner* (Toronto, Nov. 25, 1976). The authors are resource economists with the federal Dept. of Energy, Mines and Resources.

[2] For an analysis of this process see Marx, Karl, *Capital, Vol. I*, (New York, 1967), esp. p. 751.

"The discovery of gold in America, the exploitation, enslavement, and entombment in mines of the aboriginal population, the beginning of the conquest and the looting of the East Indies, the turning of Africa into a warren for the commercial hunting of black skins, signalised the rosy dawn of the era of capitalist production."

[3] Hymer, Stephen, "The Multinational Corporation and the Law of Uneven Development" in Radice, ed., *International Firms and Modern Imperialism*, (Harmondsworth, 1975), p. 49. The late Stephen Hymer's work on transnational capital is perhaps the most concise and informative available.

[4] Baran, Paul and Sweezy, Paul, *Monopoly Capital*, (New York, 1966), p. 42.

The Birth of a Monopoly

[1] Thompson, J.F. & Beasley, Norman, *For the Years to Come*, (Toronto, 1960), p. 71.

[2] Main, O.W., *The Canadian Nickel Industry: A Study in Market Control and Public Policy*, (Toronto, 1955), p. 41. Main's study is the definitive work on the development of the nickel industry in Canada.

[3] *Ibid*, p. 59.

[4] *Ibid*, p. 35.

[5] *Ibid*, p. 45.

[6] *Canadian Mining Review*, XXI, No. 4, (April, 1902).

[7] Main, *op. cit.*, p. 62.

[8] Thompson & Beasley, *op. cit.*, p. 211.

[9] *Ibid*, p. 148.
[10] Nelles, H.V., *The Politics of Development*, (Toronto, 1974), p. 329.
[11] *The Financial Post*, (Toronto, July 16, 1916).
[12] *Toronto Telegram*, (December 27, 1914).
[13] *The Daily Star*, (Toronto, August 18, 1916).
[14] *The Financial Post*, (Toronto, July 16, 1916).
[15] *Ibid*, (December 15, 1934).
[16] Main, *op. cit.*, pp. 111, 117.
[17] *Ibid*, p. 112.
[18] Hoopes, T., *The Devil and John Foster Dulles*, (Boston, 1973), p. 47.
[19] *The Financial Post*, (Toronto, May 13, 1933).
[20] Thompson & Beasley, *op. cit.*, p. 240-241.
[21] Deverell, John & the Latin American Working Group, *Falconbridge: Portrait of a Canadian Mining Multinational*, (Toronto, 1975), p. 43-46.

Who Dug the Mines?

[1] Interview, Hugh Kennedy, (Sudbury, Ontario, 1976).
[2] Interviews, Robert Carlin, (Sudbury, Ontario, 1976).
[3] Interview, Peggy Racicot, (Sudbury, Ontario, 1976).
[4] Lang, John B., *A Lion in a Den of Daniels: A History of the IUMMSW in Sudbury, Ontario, 1942-62*. (Unpublished thesis, University of Guelph, 1970). Unless otherwise noted, the following details of the history of Mine Mill are from this source.
[5] Quoted by the Western Federation of Miners, *Miners' Magazine*, Vol. IX, No. 247, (March 19, 1908).
[6] *The Sudbury Star*, (Sudbury, October 18, 1918).
[7] *Ibid*, (Sudbury, July 29, 1916).
[8] The 1937 strike against General Motors in Oshawa is the best-known of the Canadian labour struggles of this period. For details of the complex relations between the CIO in Canada, the Communist Party, and the Co-operative Commonwealth Federation, see Abella, I.M., *Nationalism, Communism, and Canadian Labour*, (Toronto, 1973).
[9] Lang, *op. cit.*, p. 31.
[10] Nelles, *op. cit.*, p. 441.
[11] Abella, *op. cit.*, p. 89.
[12] Interview, Sarah and Bill Santala, (Sudbury, Ontario, 1976).
[13] Lang, *op. cit.*, p. 42-43.
[14] Interview, Ray Stevenson, (Sudbury, Ontario, 1976).
[15] Robert Carlin, *loc. cit.* After this incident Carlin took his wife, Kay, back to Kirkland Lake and returned to Sudbury under an assumed name.
[16] *Toronto Star*, (Toronto, February 26, 1942).
[17] Peggy Racicot, *loc. cit.*
[18] Abella, *op. cit.*, p. 141.
[19] Young, Walter, *The Anatomy of a Party: the National CCF, 1932-61*, (Toronto, 1969) p. 276.
[20] Ray Stevenson, *loc. cit.*

21 *Toronto Star*, (Toronto, January 28, 1954).
22 Lang, *op. cit.*, p. 206.
23 *The Telegram*, (Toronto, November 13, 1959).
24 Interview, Albert Ouellet, (Sudbury, Ontario, 1976).
25 Peggy Racicot, *loc. cit.*
26 Interview, Patrick Bell, (Sudbury, Ontario, 1976).

Inco in Guatemala

1 Hoopes, T., *The Devil and John Foster Dulles*, (Boston, 1973), p. 137.
2 Melville, T. & M., *Guatemala — Another Vietnam?*, (Harmondsworth, 1971), p. 54.
3 North American Congress on Latin America (NACLA), *Guatemala*, (New York, 1974), p. 52.
4 Melville, *op. cit.*, p. 70.
5 Galeano, E., *Guatemala: Occupied Country*, (New York, 1969), p. 51.
6 See especially Jones, S., "Anatomy of an Intervention: The U.S. 'Liberation' of Guatemala", NACLA, *op. cit.*, p. 57.
7 Gerassi, John, *The Great Fear in Latin America*, (New York, 1965), p. 241.
8 Cited in Melville, *op. cit.*, p. 118.
9 *Ibid*, p. 51.
10 Quoted in Park, Frank & Libby, *Anatomy of Big Business*, (Toronto, 1973), p. 140.
11 Faculty of Economics, University of San Carlos, *Guatemala Contra Exmibal*, (Guatemala City, 1970). Much of the information in this section comes from this pamphlet and from Goff, Fred, "Take Another Nickel Out", NACLA, *op. cit.*, p. 151.
12 "Central American Common Market: Profits and Problems in an Integrating Economy", *Business International*, (New York, 1969), p. 27.
13 *The Globe and Mail*, (Toronto, March 2, 1971).
14 Quoted in Goff, *op. cit.*, p. 156.
15 *Financial Times*, (London, December 2, 1971). One Guatemalan official protested, "We have a constitutional clause prohibiting expropriations."
16 *Northern Miner*, (Toronto, September 18, 1975).
17 Export Development Corporation, *News Release* 74-14, (Ottawa, July 30, 1974).
18 See Cohen, I., *Regional Integration in Central America*, (Lexington, 1972), p. 60. ("As of April, 1969, (CABEI's) overall resources amounted to $250 million, of which $215 million (or 86 percent) came from foreign sources — about three-fourths from the U.S. and the Inter-American Development Bank, where the U.S. had decisive influence.") According to *Business International*, CABEI has also received medium and short-term loans from Morgan Guaranty Trust, a familiar name in Inco's history.
19 Inco, *Annual Report*, (1975).
20 Faculty of Economics, University of San Carlos, *loc. cit.*, and NACLA, *op. cit.*, p. 160.
21 *Caribbean Business News*, VI, No. 5, (February, 1976).

170 THE BIG NICKEL

[22] *Montreal Gazette*, (Montreal, August 28, 1976).
[23] *Latin America*, (London, November 26, 1976).
[24] Quoted in Lewis, E., "Guatemala: Banana Republic on the Brink of Doomsday", *Sunday Times Magazine*, (London, March 14, 1971).
[25] "Exmibal plant: getting the bugs out", *Central America Report*, IV, No. 6, (Guatemala City, February 7, 1977).

... And in Indonesia

[1] K.H.J. Clarke, assistant vice-president of Inco, in testimony to the Standing Committee on Foreign Affairs of the *United States* Senate, November 24, 1970. Mr. Clarke has acted as president of the Pacific Basin Economic Council, a business organization with representatives from the U.S., Canada, Japan, New Zealand, and Australia. See *Business Week*, (June 2, 1973).
[2] Quoted in *Pacific Research and World Empire Telegram*, Vol. I, No. 1, (August 3, 1969), p. 8.
[3] Ransom, D., "The Berkeley Mafia and the Indonesian Massacre", *Ramparts*, (October, 1970).
[4] Scott, P.D., "Exporting Military-Economic Development: America and the Overthrow of Sukarno", in M. Caldwell, ed., *Ten Years of Military Terror in Indonesia*, (Nottingham, 1975), pp. 233-234.
[5] *Ibid*, pp. 227-228.
[6] *Ibid*, p. 230.
[7] *Time Magazine*, (December 17, 1965).
[8] Quoted by Amnesty International, *Indonesia Special*, (The Netherlands, 1973), pp. 11-12.
[9] Nixon, Richard, "Asia After Vietnam", *Foreign Affairs*, (October, 1967).
[10] Appendix to "Proceedings of the Standing Senate Committee on Foreign Affairs", No. 14, (April 6, 1971), p. 16.
[11] Cited by Howell, Leon, "Indonesia: Economic Prospects and the Status of Human Rights", *International Policy Report*, II, No. 3, (Washington, December, 1976), p. 15.
[12] Payer, Cheryl, *The Debt Trap: The IMF and the Third World*, (New York, 1974).
[13] Cited in Lapizzi, E., "Trade Unions Under the New Order", the British Indonesia Committee, ed., *Repression and Exploitation in Indonesia*, (Nottingham, 1974), p. 41.
[14] *Ibid*, p. 44.
[15] *Financial Post*, (Toronto, April 19, 1975).
[16] Dodwell Marketing Consultants, *Industrial Groupings in Japan*, (Tokyo, 1973); Financial Post Corporation Service, *Inco Ltd.*, (Toronto, 1976), pp. 11-12 and *Business Week*, (March 31, 1975).
[17] Quoted in *Pacific Imperialism Notebook*, III, No. 2, (San Francisco, 1971-1972).
[18] *The Northern Miner*, (Toronto, April 24, 1975).
[19] *Ibid*, (Toronto, October 23, 1975).
[20] *CIDA Contact*, No. 43, (August, 1975).
[21] *Far Eastern Economic Review*, (Hong Kong, December 26, 1975).

[22] *The Globe and Mail*, (Toronto, August 21, 1975) and *CIDA Contact, loc. cit.*
[23] *CIDA Contact*, No. 38, (March, 1975).
[24] *Far Eastern Economic Review, loc. cit.*
[25] "Inco's Toroaks Nickel Project Dedicated by President of Indonesia", *Inco Media Information*, (Toronto, March 31, 1977).
[26] *The Globe and Mail*, (Toronto, March 11, 1976).
[27] *P.T. International Nickel Indonesia*, (Canada, no date).
[28] "Nine years and $850 million later, Inco's new plant starts up", *Australian Financial Review*, (March 30, 1977).
[29] Interview, Carmel Budiarjo, (Toronto, Ontario, 1975).
[30] "Indonesia looks to Inco when the oil runs out", *Australian Financial Review*, (February 23, 1977).
[31] Quoted in Knowles, R.S., *Indonesia Today: The Nation That Helps Itself*, (Los Angeles, 1973), pp. 85-86.
[32] Howell, *loc. cit.*, p. 15.

The Growth Strategies: The End of Motherhood

[1] *Fortune*, (New York, May, 1957).
[2] *Ibid*, (New York, January, 1975).
[3] Figures from Local 6500, USWA.
[4] "Nickel Giant Inco to Get New Chairman", *Toronto Star*, (Toronto, April 18, 1977).
[5] *Ibid*.
[6] *Financial Post*, (Toronto, March 29, 1975).
[7] *Iron Age*, (Radner, December 2, 1974).
[8] *Financial Post*, (Toronto, April 19, 1975).
[9] *Ibid*, (July 3, 1976).
[10] Clement, Wallace, *An Analysis of Economic Power*, (Toronto, 1975), p. 107.
[11] Hoopes, T., *The Devil and John Foster Dulles* ..., p. 37.
[12] Clement, *op. cit.*, p. 161.
[13] *Toronto Star*, (Toronto, May 6, 1972).

Scorched Earth ...

[1] Carter, D., and McCauley, W., *The Sudbury Pollution Problem: Socio-Economic Background*, unpublished (Department of the Environment, Ottawa, 1974), p. 94.
[2] *Ibid*, pp. 43-44.
[3] *Ibid*.
[4] *Ibid*, p. 96.
[5] *Sudbury Star*, (Sudbury, March 4, 1916).
[6] Quoted in Main, O.W., *The Canadian Nickel Industry* ..., p. 24.
[7] *Sudbury Journal*, (Sudbury, July 12, 1906).
[8] *Sudbury Star*, (Sudbury, March 4, 1916).
[9] Province of Ontario, *Damage by Fumes Arbitration Act*, (May 3, 1921), sections 3 and 5.

10 Carter and McCauley, *op. cit.*, p. 129.
11 Interview with Eli Martel, *Alternatives*, II, No. 3, (Spring, 1973), and *Inco Annual Report*, (1968), p. 19.
12 Minister of Environment, *Amending Control Order*, (Toronto, July 17, 1972).
13 Minister of Environment, *News Release*, (Toronto, January 5, 1973).
14 *Ibid.*
15 Floyd Laughren, MPP Nickel Belt, *News Release*, (Toronto, January 9, 1973).
16 Response to question no. 2816, tabled by John Rodriguez, MP Nickel Belt, (Ottawa, October 15, 1973).
17 Information from Ontario Ministry of Environment, provided by Floyd Laughren.
18 Economic Council of Canada, Annual Report, (Ottawa, 1975).
19 Interview, Keith Winterhalter, (Sudbury, Ontario, 1976).
20 *Ibid.* Winterhalter estimates the cost of this revegetation at $200 an acre.
21 Interview with Ontario Ministry of the Environment officials, (Sudbury, Ontario, 1976).
22 *Ibid.*
23 Hall, Ross H., "The Stack", *Alternatives*, II, No. 3, (Spring, 1973), p. 26.
24 *Sudbury Star*, (Sudbury, January 4, 1976).
25 Beamish, R.H., and Harvey, H.H., *Acidification of the La Cloche Mountain Lakes and Resulting Fish Mortalities*, (Fisheries Research Board of Canada, Ottawa, 1972), and Ontario Water Commission, *Preliminary Report on the Influence of Industrial Activity on the Lakes in the Sudbury Area*, (Toronto, 1969-1970).
26 Carter and McCauley, *op. cit.*, p. 57.

... And Broken Bodies

1 The International Nickel Company of Canada Ltd., *The Romance of Nickel* (1945), p. 28. Seventy-five thousand copies of this public relations pamphlet were produced in an attempt by the Company to polish its image after the organizing of a union in Sudbury.
2 "Dead of un-natural causes", *The Miners' Voice*, (Toronto, October 1977), p. 11. This article notes that one in every 754 miners died at work in the period 1966-1975.
3 G.H. Gilchrist, area supervisor for the USWA (Northeastern Ontario), *Brief to the Ham Commission*, (January, 1975).
4 *Toronto Star*, (Toronto, January 30, 1975).
5 G.H. Gilchrist, *loc. cit.*
6 *Report of the Royal Commission on the Health and Safety of Workers in Mines*, published by the Attorney-General, Province of Ontario (Toronto, 1976), pp. 138, 133. Hereafter cited as the Ham Commission.
7 *Ibid*, p. 6.
8 *Ibid*, table D7.
9 *Ibid*, table 29, p. 126.
10 Inquest Committee, Local 6500, *Report*, (April, 1975).
11 *A History of Steelworkers' Action for Occupational Health in Ontario Mining — Brief to the Ham Commission*, pp. 40-42.

¹² International Metalworkers Federation, *Health and Safety of Nickel Workers*, (Geneva, 1976).
¹³ Lowe, M., "The Worst Killer of Workers Ignored", *Last Post*, (December, 1976).
¹⁴ Ham Commission, pp. 132-133.
¹⁵ *The Globe and Mail*, (Toronto, September 10, 1976).
¹⁶ "5 dead in Thompson: Another victim of the bonus system", *The Miners' Voice*, (Toronto, July 1977), p. 4.
¹⁷ Sass, R., "The Underdevelopment of Occupational Health and Safety in Canada: Contradictions and Conventional Wisdom", in Wm. Leiss, ed., *Politics of the Environment* (forthcoming).
¹⁸ Cited in Ham Commission, p. 218.
¹⁹ *Ibid*, p. 18.
²⁰ *Ibid*, p. 40.
²¹ Williams, Lynn, Director, District 6, USWA, *Brief to the Ham Commission*.
²² International Metalworkers Federation, *op. cit*.
²³ G.H. Gilchrist, *loc. cit*.
²⁴ "Doing without the diesel", *The Miners' Voice*, (Toronto, July 1977), p. 4.
²⁵ International Metalworkers Federation, *op. cit*.
²⁶ Ham Commission, p. 230.
²⁷ *Ibid*, p. 230.
²⁸ *The Globe and Mail*, (October 7 & 8, 1976).
²⁹ "Ham Commission Timid", *Health Alert*, I, No. 1, (Toronto, September 1976).

Why Inco?

¹ Bourgault, P.L., *Innovation and the Structure of Canadian Industry*, (Science Council of Canada, Special Study No. 23, 1972), p. 126.